Entrepreneurship Fundamentals

1 2 3 4 5 6 7 8 9 LWI 27 26 25 24 23

ISBN 978-1-26-660679-3
MHID 1-26-660679-3

Cover Image: Chayanit/Shutterstock

Interior Image (abstract background):
All credits appearing on page or at the end of the book are considered to be an extension of the copyright page.

The Internet addresses listed in the text were accurate at the time of publication. The inclusion of a website does not indicate an endorsement by the authors or McGraw-Hill Education, and McGraw-Hill Education does not guarantee the accuracy of the information presented at these sites.

Meet Our Lead Content Architect

Courtesy of Kandis Spurling

Tim Broxholm (MA, Seattle Pacific University)

For many learners, their business education journey begins with Introduction to Business. Tim has had the pleasure of educating numerous unique students in this course. His innovative teaching methods, applications of learning science, and educational technology have taken this foundational course to the next level. He arms students with the skills needed to perform like pros in the world of work, excel in future college courses, build wealth, and become savvy consumers.

Tim has been teaching marketing, management, and entrepreneurship courses at Green River College in Auburn, Washington, since 2010. He was tenured in 2015 and is best known for co-creating and then leading the country's first Bachelor of Applied Science (BAS) in Marketing and Entrepreneurship. Tim also developed an Applied Management BAS program. Affordable schools.net has recognized the Marketing & Entrepreneurship program as the #1 most affordable entrepreneurship program in the country and the #7 most affordable entrepreneurship program by UniversityHQ. Tim is a recipient of Green River College's Distinguished Faculty Award.

In addition to teaching and authoring, Tim works in the technology industry as a Senior Manager at Microsoft, which provides him the ability to bridge the divide between today's higher education curriculum and the rapidly changing needs of today's employers. While Tim serves in many roles, his most important role is as husband to his wife, Danielle, and dad to his sons, Raymond and Joey.

Entrepreneurship Fundamentals would not have been possible without the generosity, innovation, and expert contributions of the following:

Content Contributors

Lisa Cherivtch, *Oakton Community College*

Helen Davis, *Jefferson Community and Technical College*

Scott Elston, *Iowa State University*

Chris Finnin, *Drexel University*

Charla Fraley, *Columbus State Community College—Columbus*

Peter Natale, *ECPI University Roanoke*

Robin James, *William Rainey Harper College*

Krissy Jones, *City University of Seattle*

Tim Rogers, *Ozarks Technical Community College*

Alice Stewart, *North Carolina A and T University*

Content Reviewers

Rebecca Adams, *Kansas State University*

Deolis Allen Jr., *Wayne County Community College*

Eric Alston, *University of Colorado—Boulder*

Lydia Anderson, *Fresno City College*

Mark Arnold, *Saint Louis University*

Wayne Ballentine, *Prairie View A and M University*

Susanne Bajt, *Harper College*

Kimberly Baker, *Bryant & Stratton College*

Laura Bantz, *McHenry County College*

Brian Bartel, *Mid-State Technical College*

Denise Barton, *Wake Technical Community College*

Michael Bento, *Owens Community College*

Gene Blackmun, *Rio Hondo College*

Curtis Blakely, *Ivy Tech Community College—Richmond*

Kathleen Borbee, *Monroe Community College*

Nicholas Bosco, *Suffolk County Community College*

David Button, *State University of New York—Canton*

Derrick Cameron, *Vance Granville Community College*

Margie Campbell Charlebois, *North Hennepin Community College*

Rafael Cardona, *Glendale College*

Janel Carrigan, *Fresno City College*

Patricia Carver, *Bellarmine University*

Carlene Cassidy, *Anne Arundel Community College—Arnold*

Lisa Cherivtch, *Oakton Community College*

Christopher Chin, *Glendale Community College*

Katherine Clyde, *Pitt Community College*

Gary M. Corona, *Florida State College at Jacksonville*

Anastasia Cortes, *Virginia Tech*

Sandy Dantone, *A-B Tech Community College*

Helen Davis, *Jefferson Community and Technical College*

Mark DeCain, *Wake Technical Community College*

Gustavo Demoner, *West Los Angeles College*

Carrie Devone, *Mott Community College*

Tina Donnelly, *Horry Georgetown Technical College*

Glenn Doolittle, Jr., *Santa Ana College*

Steve Dunphy, *Indiana University—Northwest*

Christopher Enge, *Mott Community College*

Mary Ewanechko, *Monroe Community College*

Christina Force, *Bloomsburg University of Pennsylvania*

Jill Friestad-Tate, *Des Moines Area Community College—West*

Tina Gaffey, *Horry Georgetown Technical College*

Prachi Gala, *Elon University*

Barbara Garrell, *Delaware County Community College*

Kimberly Gleason, *University of Houston—Downtown*

Joey Goodman, *Davidson County Community College*

Richard Gordon, *Illinois State University*

Francine Guice, *Oakland University*

Andrew W. Gump, *Liberty University*

Brian Gurney, *Montana State University—Billings*

Marie Halvorsen-Ganepola, *University of Arkansas*

Michele Hampton, *Cuyahoga Community College*

Stanton Heister, *Portland State University*

Austin Hill, *Harford Community College*

Linda Hoffman, *Ivy Tech Community College*

Phil Holleran, *Mitchell Community College*

Jianli Hu, *Cerritos College*

Julie Huang, *Rio Hondo College*

Miriam Huddleston, *Harford Community College*

Allison Hudson, *Miami Dade College*

Veronika Humphries, *The University of Louisiana Monroe*

Robin James, *William Rainey Harper College*

Russell Johnson, *Utah Valley University*

Gina Kabak, *Union County College*

Ahmad Karey, *Salt Lake Community College*

Eileen Kearney, *Montgomery County Community College*

Lindsay King, *James Madison University*

Greta Kishbaugh, *St. Petersburg College*

Elko Klijn, *Old Dominion University*

Mary Beth Klinger, *College of Southern Maryland*

Stephen Konrad, *Mount Hood Community College*

Jonathan Krabill, *Columbus State Community College—Columbus*

John Kurnik, *St. Petersburg College*

Eduardo Landeros, *San Diego Mesa College*

Marie Lapidus, *Harper College*

Kimberly B. Leousis, *University of Mobile*

Tammira Lucas, *Harford Community College*

Greg Luce, *Bucks County Community College*

Trina Lynch-Jackson, *Ivy Tech Community College—Lake County Campus*

Monty Lynn, *Abilene Christian University*

Anne Makanui, *North Carolina State University*

Jennifer Malfitano, *Delaware County Community College*

Marcia Marriott, *Monroe Community College*

Theresa Mastrianni, *Kingsborough Community College*

Chris McChesney, *Indian River State College*

Derine McCrory, *Mott Community College*

Donna McCubbin, *Clemson University*

Carlespie Mary Alice McKinney, *Oakland Community College*

Chris McNamara, *Finger Lakes Community College*

Ken Mullane, *Salem State University*

Jennifer Muryn, *Robert Morris University*

Shawn Myers, *Lincoln University*

Kristi Newton, *Chemeketa Community College*

Mihai Nica, *University of Central Oklahoma*

Lisa Novak, *Mott Community College*

Don Oest, *University of Colorado—Boulder*

Lois Olson, *San Diego State University—San Diego*

Hussein Othman, *Oakland Community College*

Pallab Paul, *University of Denver*

Jeffrey Penley, *Catawba Valley Community College*

Marc Postiglione, *Union County College*

Andrew Pueschel, *Ohio University—Athens*

Anthony Racka, *Oakland Community College—Auburn Hills*

Bharatendra Rai, *University of Massachusetts—Dartmouth*

Greg Rapp, *Portland Community College—Sylvania*

Chris Retzbach, *Gloucester County Institute of Technology*

Daniel Rhem, *Pitt Community College*

Raina Rutti, *Dalton State College*

Amber Ruszkowski, *Ivy Tech Community College*

Whitney Sanders, *Quinnipiac University*

Richard Sarkisian, *Camden County College*

Susan Schanne, *Eastern Michigan University*

Douglas Scott, *State College of Florida*

Raj Selladurai, *Indiana University—Northwest*

Sarah Shepler, *Ivy Tech Community College—Terre Haute*

Elizabeth Sikkenga, *Eastern Michigan University*

Amy Simon, *Jacksonville State University*

Jen Smith, *University of Idaho*

Harris Sondak, *University of Utah*

Jacob Stanford, *Chattanooga State Community College*

Alice Stewart, *North Carolina A and T University*

Theresa Strong, *Horry-Georgetown Technical College*

Mary Stucko, *Lansing Community College*

Chao Sun, *Eastern Michigan University*

Yvette Swint-Blakely, *Lansing Community College*

Jared Taunton, *Pitt Community College*

David L. Taylor, *Indiana University*

Ronda Taylor, *Ivy Tech Community College*

Jarvis Thomas, *Lone Star College*

Ron Trucks, *Jefferson College*

Gary Tucker, *Lone Star College—University Park*

Elizabeth Turnbull, *University of Alabama at Birmingham*

Mallory Tuttle, *Old Dominion University*

Alex Wadford, *Pitt Community College*

John Ward, *Rochester Institute of Technology*

Timothy Weaver, *Moorpark College*

Ann Weiss, *Ivy Tech Community College*

James Welch, *University of Tampa*

Ruth White, *Bowling Green State University*

Miriam Wiglesworth, *Harford Community College*

Irene Wilder, *Jefferson Community College*

Amy Williams, *Valdosta State University*

Deric Williams, *Wayne State University*

Luke Williams, *Central Washington University*

Mark Williams, *Community College of Baltimore County*

Shallin Williams, *Tri-County Technical College*

Rick Wills, *Illinois State University*

Elizabeth Wimer, *Syracuse University*

Ray Wimer, *Syracuse University*

Bruce Yuille, *Mid Michigan College*

Marisa Zakaria, *Glendale Community College*

Nancy Zimmerman, *Community College of Baltimore County—Cantonsville*

Symposia and Focus Group Participants

Elsa Anaya, *Palo Alto College*

Norma Anderson, *Ivy Tech Community College*

Rocky Belcher, *Sinclair Community College*

Connie Belden, *Butler Community College*

Michael Bento, *Owens Community College*

William Bettencourt, *Edmonds Community College—Washington*

Koren Borges, *University of North Florida*

Maurice Brown, *Harford Community College*

Carlene Cassidy, *Anne Arundel Community College*

Basil Chelemes, *Salt Lake Community College*

Lisa Cherivtch, *Oakton Community College*

Rachna Condos, *American River College*

Debra Crumpton, *Sacramento City College*

Tyler Custis, *University of South Dakota*

Gustavo Demoner, *West Los Angeles College*

John DeSpagna, *Nassau Community College*

Tina Donnelly, *Horry Georgetown Technical College*

Erik Ford, *University of Oregon*

Tracy Fulce, *Oakton Community College*

Kathleen Gallagher, *Johnson County Community College*

Wayne Gawlik, *Joliet Junior College*

Kimberly Gleason, *University of Houston—Downtown*

Shari Goldstein, *San Jacinto College*

Mark Grooms, *Orange Coast College*

Michele Hampton, *Cuyahoga Community College*

Phil Holleran, *Mitchell Community College*

Julie Huang, *Rio Hondo College*

Miriam Huddleston, *Harford Community College*

Samira Hussein, *Johnson County Community College*

Ralph Jagodka, *Mt. San Antonio College*

Robin James, *William Rainey Harper College*

Jonathan Krabill, *Columbus State Community College—Columbus*

Janet Kriege-Baker, *Ivy Tech Community College*

Steven Levine, *Nassau Community College*

Catalin Macarie, *University of Cincinnati*

Theresa Mastrianni, *Kingsborough Community College*

Paulette McCarty, *Northeastern University*

Kimberly Mencken, *Baylor University*

Steven Mohler, *Ivy Tech Community College*

Peter Natale, *ECPI University Roanoke*

Mihai Nica, *University of Central Oklahoma*

John Nobles, *Ohio University*

Margaret O'Connor, *Bloomsburg University of Pennsylvania*

Kathy Osburn, *Antelope Valley College*

Andrew Pueschel, *Ohio University*

Ujvala Rajadhyaksha, *Governors State University*

Steve Riczo, *Kent State University—Kent*

Donald Roomes, *Florida International University—Miami*

Carol Rowey, *Community College Rhode Island—Lincoln*

John Russo, *Irvine Valley College*

Raina Rutti, *Dalton State College*

Mark Ryan, *Hawkeye Community College*

Amy Santos, *State College of Florida—Manatee*

Steven Sedky, *Santa Monica Community College*

Phyllis Shafer, *Brookdale Community College*

Carl Smalls, *Guilford Technical Community College*

Nayrie Smith, *Miami Dade College*

Ray Snyder, *Trident Technical College*

Jeff Spector, *Middlesex County College*

Alice Stewart, *North Carolina A and T University*

Theresa Strong, *Horry Georgetown Technical College*

Ronda Taylor, *Ivy Tech Community College*

George Valcho, *Bossier Parish Community College*

Jacob Voegel, *Coastal Carolina University*

Ruth White, *Bowling Green State University*

Rick Wills, *Illinois State University*

Doug Wilson, *University of Oregon*

Mark Zorn, *Butler County Community College*

Board of Advisors

A special thank you goes out to our Board of Advisors:

Koren Borges, *University of North Florida*

Elisabeth Cason, *Bossier Parish Community College*

Gustavo Demoner, *West Los Angeles College*

Katie Gallagher, *Johnson County Community College*

Miriam Huddleston, *Harford Community College*

Greg Luce, *Bucks County Community College*

Kimberly Mencken, *Baylor University*

Phyllis Shafer, *Brookdale Community College*

Chapter 6: Understanding Your Customer

Chapter 7: Accounting and Financial Statements

What To Expect

By the end of the chapter, you will be able to:

- Summarize the concept of entrepreneurship.
- Describe the characteristics associated with being an entrepreneur.
- Summarize different forms of entrepreneurship.

Copyright © McGraw Hill ESB Basic/Shutterstock

ARTistic: Art for Everyone

"A Problem Is an Opportunity, a Spark of Innovation that Leads to an Idea for a Whole New Business . . . but, Where Do I Begin?"

Carl, who has always loved art, has been working at a trendy gallery in Chicago for the past four years. He loves assisting the gallery's director in finding and displaying works by great new talent.

However, Carl notices that a number of his friends and family believe they cannot afford the art in his gallery, and a number of the artists he meets struggle to get into the limited number of galleries impacting their ability to generate enough income forcing them to be the cliché "starving artist."

Carl keeps asking himself. "Why can't art be affordable to every consumer?" He also wonders why isn't there a platform dedicated to help artists sell their works directly to consumers and get artists to more galleries. He thinks, this is a problem that needs a solution—and the idea for ARTistic.com is born.

As Carl envisions his business, ARTistic will be an online marketplace that allows artists to sell their artwork directly to customers. The site also connects artists, art lovers, and gallery curators which Carl believes can help democratize art for all.

He is discussing his idea with his Aunt Samantha, who started her career as a software engineer and is now a technology executive who has experience building marketplace platforms.

I think your idea has tremendous potential, Carl. I checked the domain directory, and I found an available website. I went ahead and bought it for you.

Wow—Thanks, Aunt Samantha! There are other websites that sell art by thousands of different artists, like Etsy and ArtWeb, ARTistic is going to put the emphasis on building community and affordability. My homepage is going to say "Art for All."

I always thought you loved working at the gallery. Why did you decide to strike out on your own?

Everything at the gallery is so expensive. I have the passion and experience in selling art and based on my research I see the market potential in a relatively untapped market.

You're a true entrepreneur. And once you start ARTistic, you'll be a business owner.

That's where I need help. I know a lot about art, but not much about starting and running a business.

Carl is an entrepreneur. However, he has a lot to learn about business start-ups before he launches the ARTistic website.

In this lesson, we will discuss concept of entrepreneurship. We'll also examine the key characteristics of successful entrepreneurs. Finally, we'll discuss different forms beyond traditional entrepreneurship

Understanding Entrepreneurship and the Different Forms of Entrepreneurship

The Essence of Entrepreneurship

Carl continues his conversation about his idea for a business with his Aunt Samantha.

What Is Entrepreneurship?

The process of starting up a business is referred to as *entrepreneurship*. An *entrepreneur* is a person who sees a new opportunity for a product or service and who risks time and money to start a business with the goal of making a profit.

Entrepreneurship involves taking risks to create a product or service that fills a consumer need or opportunity in the marketplace, and subsequently building a business around the offering.

Entrepreneurs often risk their time, personal assets, resources, and money which may include personal funds, funds of investors, and even borrowed money. Individuals build their chances of success when they can use critical thinking, creativity, and problem-solving to create a strategy, and then execute efficiently on their plans.

So are you planning to hire people when you start ARTistic? Or are you going to do everything yourself?

I have a very limited budget, so initially I expect I'm going to be doing as much as I can to keep my expenses to a minimum.

You're not alone. Entrepreneurs often say that they work all day, everyday, but the rewards are significant once the business accelerates.

Entrepreneurial business owners follow a process of:

1. **identify and qualify a need in the marketplace**
2. **quantify the size of the market opportunity**
3. **translate the opportunity into a new or revised product or service**
4. **start or grow a firm**

Some entrepreneurs hope to grow their business substantially over time while others choose to remain small, however both paths require profitability to have sustainability in the marketplace.

The entrepreneurial spirit is expressed in many ways. However, as the dream and passion is pursued, the true entrepreneur "always searches for change, responds to it, and exploits it as an opportunity," according to the late management professor Peter Drucker.

Why Do People Become Entrepreneurs?

People like Carl usually become entrepreneurs for two different reasons—opportunity and necessity:

- Many entrepreneurs are ***opportunity entrepreneurs***. They are ambitious and start a business to pursue an opportunity (and profits).
- Others are ***necessity entrepreneurs***, people who suddenly must earn a living and are simply trying to replace lost income.

What do you think is motivating you to become an entrepreneur, Carl?

I really think ARTistic can fulfill an opportunity in the marketplace for both artists and consumers.

New and aspiring artists need a relatively low-cost way to promote their art and build their brand.

Small and new galleries also need an opportunity to promote their spaces to a larger audience, and many art lovers, who want to purchase pieces can't afford traditional gallery prices.

Based on my experience working with artists and managing the gallery, along with my passion for art and connecting customers and artists I believe ARTistic could be a profitable business.

Decisions

1. As Carl starts ARTistic, he is considered an entrepreneur.

   A. False

   B. True

2. Which statement best explains why Carl has chosen to start his own business?

   A. Carl is an opportunity entrepreneur.

   B. Carl is a necessity entrepreneur.

Correct Answers: 1. B; 2. A

What Does It Take to Be a Successful Entrepreneur?

While having the right business idea and executing on that idea at right time is essential to be a successful entrepreneur developing some additional skills and abilities aid in entrepreneurial success.

- **Critical Thinking:** Entrepreneurs who engage in critical thinking involves deep and continual questioning and analysis. When engaging in critical thinking an entrepreneurs find themselves listening to diverse experiences and perspectives, they spend time learning from valid, reliable sources of information, and They also consistently carve out time for deep thought, analysis and humbly challenge their own thought process.

- **Problem Solving:** While a number of entrepreneurs are glorified for their breakthrough innovations another key element to being a successful entrepreneur is problem solving. Entrepreneurs use a variety of approaches to try and solve problems, including brainstorming, experimentation, testing, user observations, and root cause analysis to come up with a solution.

- **Creativity and Innovative Thinking:** Entrepreneurs spend a lot of time generating ideas, the volume of their idea creation is often what lead to a creative or innovative solution. The creative process takes time, so successful entrepreneurs take the time to *ideate* on ways to create better products for customers, enhance quality of service for clients, and find ways to improve workplace for employees.

- **Time and Energy Management:** Entrepreneurs play a wide variety of roles in the early stages of the business. Primarily, those roles focus on strategy and execution. Therefore, an entrepreneur needs to be able to effectively and efficiently manage their time and energy.

- **and Professional Selling:** Entrepreneurs must sell their idea, product, and service to stakeholder and customers. Therefore, it is important to develop personal and professional selling skills.

Many entrepreneurs cultivate key psychological characteristics that make them different from those who want to work for an established business.

Self-Confidence and Belief in Personal Control

Entrepreneurs need to be self-confident and to think they have personal control of their destinies. Entrepreneurs also need confidence to act decisively.

Need for Achievement and Action Orientation

In a company, employees are often motivated by promotions, rewards, raise perks, and power.

Entrepreneurs are often more motivated to execute their ideas and realize financial rewards.

They are also action-oriented, wanting to get things done as quickly as possible rather than going through necessary established protocols, bureaucracy, and standardized processes.

Tolerance for Ambiguity and Risk

While employees in a company need to be able to make decisions based on unclear or incomplete information entrepreneurs must have even more tolerance for ambiguity because they are trying things they haven't done before.

Entrepreneurs also need to be willing to take risks, even personal financial risks, in the pursuit of new opportunities.

Energy

Inside an organization individuals may have to put in long hours to execute their defined role in the organization. As a result, the employee receives compensation, benefits, and other potential rewards such as advancement for their performance.

Entrepreneurs also invest a great deal of time and energy in building their business, often playing a variety of roles. Additionally, their investment of time and energy offers limited financial security, which creates an additional level of stress and requires a level of grit and perseverance to push through in hopes of creating a sustainable business.

Take Time to Reflect on the Road Ahead

Prior to starting an entrepreneurial journey, individuals will want to spend some time exploring and reflecting on why they want to become an entrepreneur. A couple of questions to explore include:

1. *What is your "Why"?* reflect on the reasons you want to pursue entrepreneurship with focus on your purpose, contribution, and impact.

2. *Am I financially ready for the journey?* By assessing this question, an entrepreneur can determine their willingness and ability to absorb the financial stress that comes with entrepreneurship. By reflecting on the financial question

an entrepreneur can also think strategically about the speed and intensity they may need or want to put into their business. An entrepreneur can also engage in worst-case scenario planning such as moving in with friends or family.

The Many Forms of Entrepreneurship

The risk associated with the "all-in" traditional concept of entrepreneurship may preclude an individual from believing they could be an entrepreneur. However, there are many forms and degrees of entrepreneurship.

The following are additional opportunities for an individual to pursue their own entrepreneurial aspirations.

Intrapreneur

An *intrapreneur* works inside an existing organization, sees an opportunity for a product or service, and mobilizes the organization's resources to turn the opportunity into a profitable reality.

The intrapreneur might be a researcher, marketer, or scientist, but could also be an employee who sees an opportunity to create or offer a new product or service to customers.

The iconic McDonalds Happy Meal is an example of intrapreneurship, as it was developed by regional manager who wanted to create an offering aimed at children.

Entrepreneurial Team

An *entrepreneurial team* is a group of people with different kinds of expertise who form a team to create a new product.

One variant is a so-called *skunkworks*, a team whose members are separated from an organization's normal operation and asked to produce a new, innovative project.

For example, during World War II, defense contractor Lockheed Martin, assembled a secret team to build a fighter jet in 150 days that would help defeat the Germans. Today, it is common for technology companies to create their own secret teams that work on future innovations.

Copyright © McGraw Hill   Milan Ilic Photographer/Shutterstock

Micropreneur

A *micropreneur* takes the risk of starting and managing a business that remains small (often home-based). This allows them to do the kind of work they want to do and may offer a more balanced lifestyle.

For example, an individual enjoys helping people find new careers may start a career coaching practice that allows them to have flexibility and autonomy while generating income from home.

Solopreneur

A *solopreneur* is a business owner who owns and operates his or her business alone.

For example, a person may source products from a manufacturer and then sell them on sites like Amazon or eBay. The individual manages all the business activities by themselves and uses technology tools and platforms to run the business alone.

Side Hustle

A myth of entrepreneurship is that an individual simply quits their day job and immediately launches a new business. The reality is most entrepreneurs start with a *side hustle*, which involves working on a project or business idea outside of one's regular employment.

This form of entrepreneurship allows an individual to pursue an idea, interest, or business concept while not forgoing the security of their day job. A side hustle can also create a new source of income and may morph into a full-time business. It is important to note that an individual needs to refer their companies polices to ensure no rules or regulations are being violated.

For example, if a person with a passion for pets and an interest in baking might create their own dog treats and sell them via an online store or at local farmers markets on the weekends.

Freelancer

Another form of entrepreneurship is to become a *freelancer*. Freelancing is a form of self-employment in which a person leverages their knowledge, skills, and abilities to perform tasks, services, or functions for a customer.

For example, if an individual has the ability to create engaging content, they could freelance as a social media consultant that creates content for companies looking to outsource their social media marketing function. Websites such as Fiverr, Guru, and Upwork provide a platform for freelancers and potential clients to connect.

The Start-Up Process

By the end of this lesson, you will be able to:

- Explain the rationale behind creating a business.
- Explain the steps an entrepreneur should take before writing a business plan.
- Outline the elements of a traditional business plan.

A Plan to Launch ARTistic

"How Do I Organize My Plans and My Needs for My New Business?"

Carl's Aunt Samantha, an executive a technology company, has told him that an organization exists within the U.S. federal government to help entrepreneurs: the Small Business Administration (SBA). Because Carl wants as much good advice as he can get, he makes an appointment with an SBA mentor, Tina.

I'm glad you came to us, Carl. Can you share with me where you are at with your business idea?

Tina, at this point, everything is mostly in my head.

Let me encourage you to put your plans in writing by creating a pitch deck and a business plan. It will help you with your thinking.

The pitch deck can be an asset as you meet with potential investors or collaborators, and your business plan will be essential if you try to secure a loan through a bank.

I was reading your mission statement for ARTistic, and I liked where you said your mission is "Art for all."

Yes, that's my goal: to create a community and marketplace that brings together artists with an untapped audience of potential buyers.

That seems like a solid idea. But before you invest large amounts of time and money, you need to think deeply about your business, competitive advantages, customers, and profit potential.

I'm hoping you can help me with all of that.

Carl has made a wise choice as he thinks about starting up ARTistic: He has asked the advice of successful businesspeople. Carl is smart enough to know what he *doesn't* know. Carl learns that his first step is to assess his overall business concept.

In this lesson, we'll examine how entrepreneurs come up with an idea for a new business solutions and how they learn about the market. We'll also outline the basic elements of a business plan, a key document that guides small business owners as they prepare for opening day (and beyond).

Getting Started: Problems and Solutions

Identifying problems are the foundation of being an entrepreneur. Prior to confronting a problem, an entrepreneur strives to find and clarify its root cause.

Entrepreneurs will spend time observing the problem, talking with those impacted by the challenge, and reviewing the problem with peers to ensure that it's a large enough problem to invest the time, money, and resources in solving.

A Better Solution Instead of a New Solution

When individuals think of the solutions created by entrepreneurs, they often think of the high impact solutions like the automobile, computer, or cell phone. These thoughts lead some to believe they must innovate or create something brand new or radically different to be a successful entrepreneur.

However, this is simply not the case. In fact, there are many additional problems consumers face that have solutions that simply need to be deployed or improved upon by an individual willing to take action.

The following are additional opportunities to solve a problem faced by consumers without having to invent something novel:

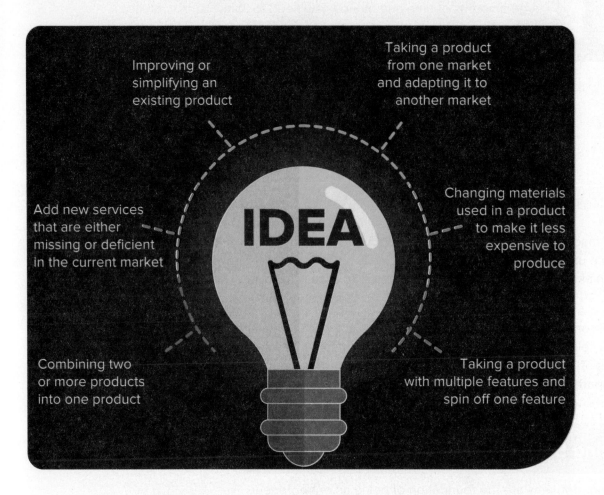

Improving or simplifying an existing product

Taking a product from one market and adapting it to another market

Add new services that are either missing or deficient in the current market

Changing materials used in a product to make it less expensive to produce

IDEA

Combining two or more products into one product

Taking a product with multiple features and spin off one feature

Pre-Start-Up Planning: Quantitative and Qualitative Analysis of Products and Services

When evaluating the opportunity for a company's products and services in the marketplace research is needed. The balance of numbers (quantitative) and satisfaction to humans (qualitative) are needed.

- **Quantitative research and evaluation** allows a business to leverage mathematical analytics and supporting data to inform decision makers with statistics about, for example, the size of a market for a specific product. The collected data can assist with understanding the number of interested buyers in a certain geographic area, age group, or income bracket.

- **Qualitative research and evaluation** involves gathering feedback, anecdotes, and sentiment from people. Often times, this includes learning about consumers' perceptions, attitudes, habits, and feelings about products, services, or brands.

This dimension of research helps to have balance for decision making in that it creates a picture of opinions. The qualitative data is often used to gain insights that help with the problem-solving aspect of innovation when contemplating developing a new product or service.

Is the Solution to the Problem Worth Solving?

While an individual may identify a problem and solution, that does not automatically translate to a profitable business. Additionally, an entrepreneur should consider:

- Does the entrepreneur have interest, passion, and commitment to solve the problem?
- Are consumers willing to pay for the solution, and if so, is the price consumers are willing to pay enough to build a sustainable business?
- Is the market opportunity large enough?
- Is the product or service unique and differentiated enough compared to other established offerings easily available in the market?
- Is the entrepreneur willing to invest time, money, and resources to solve the problem?

How Do You Learn about Your Business and the Marketplace?

Research is essential to starting any small business. To learn about the market that a business will be competing in, an entrepreneur can follow the three key steps outlined below.

Step 1: Read about Business (in General) and Your Prospective Business

Many businesspeople consider *The Wall Street Journal*, *The Financial Times*, and *Business Week* required reading for understanding the contemporary marketplace. In addition, each industry has numerous websites where people write and talk about the industry.

Step 2: Talk to Knowledgeable People

A **mentor** is an experienced person who coaches and guides lesser-experienced people by helping them understand an industry or organization's culture and structure. Usually we think of mentors as supervisors or experienced people in an organization with which we are newly employed. However, it's possible to get a mentor to help you understand the workings of the new business you're trying to start.

The Small Business Administration offers mentors through its **SCORE** program—the **Service Corps of Retired Executives**, consisting of retired executives who volunteer as consultants to advise small businesspeople. SCORE advisors can, for example, help you figure out a five-year financial plan for your business. SCORE is supplemented by the **Active Corps of Executives (ACE)**, which is composed of executives who are still active in the business world but have volunteered their time and talents.

A **trade association** consists of individuals and companies in a specific business or industry organized to promote common interests. A trade association can provide research data, training, certification programs, and links to others in your prospective field.

Step 3: Get Experience

There is no substitute for experience, which is rightfully called "the greatest teacher." You can't build custom furniture if you've never taken a carpentry course. You can't start a sales business if you've never sold anything.

Decisions

Carl once again calls his Aunt Samantha and asks her to share her experience with him. As you read their conversation, choose which term best completes the sentence.

I'm delighted to share my expertise and experience with you, Carl. I want your business to succeed.

1. Thanks, Aunt Samantha, for being my _____.

   A. SCORE

   B. funding source

   C. mentor

   D. ACE

2. After Carl spoke with Aunt Samantha, he plans to become a member of all of the following groups, which are composed of people who work in the art industry: the Society of Illustrators, Artists and Designers (SAID); the Graphics Artists Guild; and the Association for Creative Industries. All of these are examples of _____.

   A. skunkworks

   B. lobbying groups

   C. SBA advisors

   D. trade associations

Correct Answers: 1. C; 2. D

Business Plans and Pitch Decks

Carl meets with Aunt Samantha for some advice.

What Is a Business Plan?

A **business plan** is a document that outlines a proposed firm's goals, the methods for achieving them, and the standards for measuring success.

There are two important reasons to write a business plan.

Creating a business plan helps you get financing: A formal business plan shows lenders and investors what they want to see before they invest in your startup.

When you ask investors to help fund your business you need to provide details that show the roadmap for return on their investment. Remember, entrepreneurs must always be good stewards of the resources they have, whether it's investor capital or internal funds from an employer.

Creating a business plan helps you think through important details: In most cases, business ideas shouldn't be rushed to market. So, entrepreneurs take the necessary time to research and write a thorough business plan that clearly identifies the problem, solution, size of the market, and the amount of funding needed to develop and bring to market the product or service.

By realistically addressing the businesses offerings, competition, financials, and go-to-market strategy investors and lenders can gain a better understanding of their potential return on investment.

Now that you've thought about your business idea, it's time to put everything down on paper.

You mean just write my ideas down?

No, I mean a more formal business plan that will help possible investors and lenders see how you plan to run the business and earn a profit.

That said, some experts believe that a business plan is not always necessary. The arguments:

1. *You may need to act fast.* By the time you spend months writing a 60-page business plan, your opportunity may be gone.

2. *You may be able to get financing without a formal business plan.* You might be able to pitch investors with an informal 5- or 10-page summary or a PowerPoint presentation often called a *pitch deck*.

3. *You may want to protect your key strategic and competitive information.* Many entrepreneurs keep their secrets close to the vest to prevent others from stealing them.

Still, most experts agree that a business plan is important especially if an entrepreneur is seeking a loan from a bank or the SBA.

How to Engage in Research for a Business Plan

Investors will want information about your **business model**—that is, the needs the firm will meet, the operations of the business, the company's components and functions, and its expected revenues and expenses. A business plan provides all this information.

The first step in writing a business plan is researching answers to the following questions.

Business Plan Questions	
Key Question	**Additional Questions**
What's the basic idea behind my business?	How would you explain your proposed business to someone? What needs does the business meet?
What kind of industry am I entering, and how is my idea different?	Is the industry highly competitive? How will your business differ from your competitors?
How will I market to customers?	Who are your customers, how will you market to them, and what will you charge?
What qualifies me to run this business?	How does your background qualify you to start and operate this business?
How will I finance the business?	Will you be getting loans from your family, or will you be financing your business with savings, credit cards, or bank loans? What kind of returns do you plan to give investors?
When will a potential return on the investment be realized?	Investors and lenders will need to have an understanding of when and at what level of increase their investment will earn.

Analyze Potential Competition

A key goal of conducting competitive evaluations is to develop a clear understanding of the other companies that sell their offerings in a particular marketplace. By conducting competitive evaluations, an entrepreneur can:

- minimize risk as organizations attempt to compete more aggressively, scale their sales, or enter into a new market
- refine strategies by understanding the strengths and weaknesses of competitors and examine the benefits of products and services that may have had several rollouts in the market
- potentially turn the competition into a strategic partner by filling gaps in each other's weaknesses

Components of the Business Plan

Business plans usually include the following seven components.

Executive Summary

This 1- to 10-page section (typically written last) provides a short description of your product or service, its market potential, and what distinguishes it from competitors. This is the place to really sell your idea to your reader.

Introduction

This section describes the general nature of the industry you propose to compete in, the product or service you will offer, its unique characteristics, and why customers will want and benefit from it.

Marketing

Here an entrepreneur describes the market they are going into; competitors; strategy for identifying, contacting, and servicing customers; advertising plans; pricing; and projected sales. If the venture is a retail business, provide an analysis of the store's location, including vehicle and pedestrian traffic patterns and the ages and socioeconomic status of projected customers.

Manufacturing

If the business is manufacturing a product, this section should describe how and where the product will be made, the machinery and labor required, and inventory and quality-control methods.

Management

This section of the business plan discusses the background and qualifications of the owners and others who will manage the organization. If the plan can't show that the team has the necessary expertise, then consultants or other experts need to be identified as individuals who can be called upon for help.

The goal is to describe the firm's directors, other investors, management style, and compensation proposed for executives.

Legal

Here the plan should discuss the legal plan for the business, whether sole proprietorship, partnership, corporation, or some other form of legal organization. The plan should also describe the licenses, permits, and insurance that are needed and name the attorney who will handle any potential legal matters.

Financing

In this section, the plan gets into the numbers: expected revenue and expenses every year for the first five years. The plan outlines month-by-month cash flow for the first two years, and when the company expects to *breakeven*.

The plan should state what funds are need now and will need in the future. The goal is to offer best-case and worst-case scenarios for each of these, and avoid rosy promises.

Creating a Pitch Deck

In addition to building a business plan, entrepreneurs should consider creating a pitch deck. A pitch deck serves as a more succinct version of the business plan that can be presented and shared with investors.

While a business plan is a deep dive into the business and demonstrates thoughtful analysis, a pitch deck communicates to investors the story and opportunity of the business in a short yet meaningful way.

A pitch deck is typically comprised of the following components:

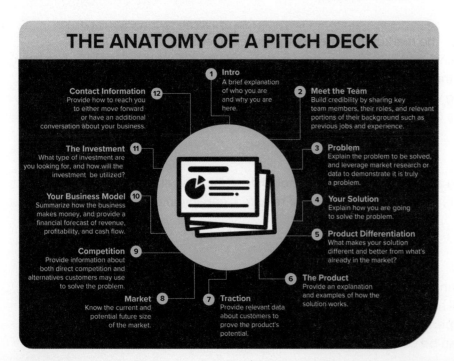

THE ANATOMY OF A PITCH DECK

1 Intro
A brief explanation of who you are and why you are here.

2 Meet the Team
Build credibility by sharing key team members, their roles, and relevant portions of their background such as previous jobs and experience.

3 Problem
Explain the problem to be solved, and leverage market research or data to demonstrate it is truly a problem.

4 Your Solution
Explain how you are going to solve the problem.

5 Product Differentiation
What makes your solution different and better from what's already in the market?

6 The Product
Provide an explanation and examples of how the solution works.

7 Traction
Provide relevant data about customers to prove the product's potential.

8 Market
Know the current and potential future size of the market.

9 Competition
Provide information about both direct competition and alternatives customers may use to solve the problem.

10 Your Business Model
Summarize how the business makes money, and provide a financial forecast of revenue, profitability, and cash flow.

11 The Investment
What type of investment are you looking for, and how will the investment be utilized?

12 Contact Information
Provide how to reach you to either move forward or have an additional conversation about your business.

Decisions

1. In assessing his business idea, Carl decides that ARTistic will be a web-based e-commerce site that generates revenue by charging a commission to artists who sell their art on the site. His expenses will be minimal because the ARTistic's storefront is virtual. Carl has just summarized his _____.

   A. business plan

   B. competitive advantage

   C. marketing plan

   D. business model

2. Choose the portion of the business plan that best characterizes this example from Carl's business plan. The ARTistic website will bring together hundreds of artists who sell their unique, handmade art at prices that the average person can afford (a clear benefit not only for the art-loving public, but also for the artists who sell their work on ARTistic).

   A. Legal

   B. Marketing

   C. Manufacturing

   D. Introduction

3. Choose the portion of the business plan that best characterizes each example from Carl's business plan.
   I will purchase physical and e-mail mailing lists to reach art lovers. I will also purchase Web ads offering first-time buyers a 10% discount, and I will work with other companies in a "link exchange" in which I link to other art-related businesses on the ARTistic website. These other businesses will then return the favor.

   A. Executive Summary

   B. Manufacturing

   C. Marketing

   D. Legal

4. Choose the portion of the business plan that best characterizes each example from Carl's business plan.
   ARTistic is essentially a portal that allows art lovers to view and purchase art directly from the artists who created it. ARTistic therefore has no manufacturing costs; most of its expenses will be advertising- and marketing-related.

   A. Financing

   B. Marketing

   C. Manufacturing

   D. Legal

5. Choose the portion of the business plan that best characterizes each example from Carl's business plan.

ARTistic.com will require quite a lot of business contracts. Each of the artists will need to sign a contract that spells out key financial details and responsibilities. For example, the artists are responsible for shipping the art to customers. I will collect the payments, deduct a 15% commission for myself, and then pay each artist quarterly.

A. Executive Summary

B. Legal

C. Financing

D. Introduction

Correct Answers: 1. D; 2. D; 3. C; 4. C; 5. B

Financing a Start-Up

By the end of this lesson, you will be able to:

- Explain the different sources of potential funding for a business idea.
- Describe the resources available to an entrepreneur who is starting a business.

Raising the Funds to Start ARTistic

"I Need Money to Start My Business. Where Do I Go?"

Carl is convinced he has a great business idea, and he has done the necessary research to build a comprehensive business plan that support his confidence.

However, he is unsure of how to fund his business and is worried about having enough money to build the website, pay for marketing campaigns and the keywords needed for search engines to find ARTistic.

He also wants to ensure that he can and keep the business operations afloat during its first year. Carl is discussing the situation with his SBA consultant, Tina.

> Carl, you mentioned that you've pretty much spent all your savings.

> Tina, yes, the expenses racked up more quickly than I expected.
>
> I thought that getting a website together would be relatively inexpensive. But building our site meant investing in seamless integration with an ordering, payment, and shipping system, which became complicated and expensive.
>
> However, getting that platform correct is essential for our business's success.

Not skimping on the website was a good decision on your part. Your website is a major part of your brand.

But now my key question is: Where do I go now that I am running low on funds? I have to borrow money or get investors, but I don't know where to start.

The good news is there are plenty of options to find funding. The best next step is for you to think about your goals and identify what type of funding makes the most sense for you and the business.

Carl is facing a challenge that many entrepreneurs face: Starting a business is costly and often not all the expenses are identified in the initial business plan.

"You have to spend money to make money," so the saying goes—but what sources of funding are available to Carl? How can he raise the capital he needs to both launch ARTistic and meet his strategic growth goals?

In this lesson, we will describe the eight sources of funding for new businesses.

By understanding the different sources of funding available, Carl will be able to decide which ones best fit his business plan and his short- and long-term goals. We will also explain how incubators help businesses grow and how enterprise zones support business.

Finding Capital to Launch a Business

Carl is wondering about sources of funds to help fund his new enterprise.

Similar to financing an established business, most entrepreneurs can access capital through borrowing money in the form of loans, or an entrepreneur can trade ownership in the company called *equity* for a specified amount of capital.

When determining the type of financing they want to pursue, an entrepreneur often involves asks:

- How much debt is the entrepreneur and the business willing or able to take on?

- Is the entrepreneur willing to take on a partner or give up some ownership in the business? If so, how much ownership?
- Will the company need additional funds in the future? If so, how does that impact the value of the equity?

For most entrepreneurs, there are typically these possibilities for financing.

Personal Savings, Credit Cards, and Second Mortgages

Entrepreneurs often fund their new enterprises by *bootstrapping* their start-up by using their personal savings or, alternatively, by maxing out their credit cards, taking a second mortgage on their homes, or taking money out of their retirement plans.

These methods are a gamble; if the business doesn't succeed, then the entrepreneur may take a severe financial hit.

Still, it is important to know that lenders often insist that you put up a sizable sum yourself, such as 30% of start-up costs, as proof of your commitment to the new business. So, in a sense, there is an important benefit of self-funding: When you're spending your own money, you're much less likely to waste it.

Family and Friends

Two other popular sources of funds for entrepreneurs—indeed, for 90% of all start-ups—are family and friends. Family members can range from those in the immediate family to an extended chain of relatives. A mom-and-pop business may often be financed by mom and pop themselves.

You should document in writing any loans that you take from family and friends. The document should state how much you are borrowing, what interest you will pay, and when you will pay back the loan in full. Having such a document will make loans from family and friends seem as "real" as loans from a bank.

Supplier and Barter Arrangements

Suppliers can help with financing your small business in three ways:

- **Short-term credit:** A supplier might provide you with short-term credit, giving you 30 to 45 days to pay for supplies or services.
- **Extended-payment plans:** When you purchase heavy-duty machinery, such as trucks and mainframe computers, the supplier may give you an extended-payment arrangement, allowing you (say) three to five years to pay for your purchases. Of course, you will also pay interest to the companies offering the extended-payment plans.
- **Barter arrangements:** To *barter* is to trade goods or services without the exchange of money. For example, you might trade your old Ford F-150 pickup to a roofer for replacing your restaurant's roof.

Loans from Financial Institutions

Financial institutions include banks, savings and loans, and credit unions.

Although normally reluctant to make loans to start-ups, financial institutions may do so if you have a good credit history or can offer some *collateral*, which is property (such as your house) to secure the loan.

During tough economic times, even good customers can find it hard to get loans from banks.

Best advice: Get to know some bank loan officers and try to educate them about your business.

Small Business Administration–Backed Loans

If you don't qualify for a loan with your bank, you might be able to get a bank loan that is guaranteed by the **Small Business Administration (SBA)**, the principal U.S. government agency charged with aiding small businesses by providing help in financing, management training, and support in securing government contracts.

The SBA has about 1,000 Small Business Development Centers across the United States, which provide free, in-depth counseling.

Although the SBA doesn't usually make direct loans, it guarantees a great many loans made to small businesses. The Small Business Investment Company (SBIC) program is a public/private partnership designed to give small businesses access to capital.

Angel Investors and Venture Capitalists

Individuals who invest their own money in a private company, typically a start-up, are known as **angel investors**.

These are people who already have a lot of money and want to make more by getting into a business with high growth potential.

Have you ever seen the TV show *Shark Tank*? The businesspeople who serve as "sharks" are considered angel investors who are willing to take a chance on start-up businesses with potential.

So is Marcus Lemonis, who serves as an angel investor for worthy companies profiled on the TV show *The Profit*.

Venture capitalists are generally companies, not individuals, that invest in new enterprises in return for part ownership of them—perhaps as much as 60% ownership.

Venture capital firms blossomed in California's Silicon Valley during the dot.com boom of the late 1990s.

However, they provide very little financing for most small businesses, being drawn mainly to technology, medical devices, biotech, green technology, and Internet-specific enterprises that have the potential to be valued for millions or billions of dollars.

Public Stock Offerings

The expression *going public* means that a privately owned company becomes a publicly owned company by issuing stock for sale to the public.

A public stock offering would not be a method of financing a start-up, but a small company that has attained a substantial measure of success might later sell stock as a way of raising the funds needed to grow the business.

Public stock offerings are usually the means by which angel investors and venture capitalists hope their early investment will result in earning them a fortune (which is exactly what happened to the early backers of Yahoo! and Google).

Micro Loans, Peer-to-Peer Lending, and Crowd Funding

There are many other sources of funds for new businesses.

- Others obtain small loans, "*microloans*" of $500 to $35,000, from organizations such as the Women's Initiative for Self-Employment.
- *Peer-to-peer (P2P) lending* involves borrowing money from individuals instead of a traditional financial institution like a bank or credit union. For example, a site like Prosper.com serves as a peer-to-peer lending group.
- *Crowdfunding* on sites like Kickstarter allows entrepreneurs to raise funding for a project from a large number of individuals.

Community Development Financial Institutions Fund

A *Community Development Financial Institution (CDFI)* is a private financial institution who provide investing as well as personal and business lending opportunities to underserved communities within the United States.

Often because of their focus on social impact these institutions receive federal funding from the United State Department of Treasury. CDFI's can also receive funding from the private sector.

Decisions

Carl is brainstorming potential opportunities to raise capital for ARTistic. Based on Carl's idea, choose the appropriate form of capital.

1. A web designer agrees to design a certain section of the ARTistic website for free if Carl sells the web designer's art at no commission on ARTistic.

   A. barter arrangement

   B. financial institution

   C. second mortgage

   D. family and friends

2. Carl's parents have agreed to lend him $10,000, and Carl has given his parents an IOU for that amount.

    A. venture capitalists

    B. public stock offering

    C. family and friends

    D. angel investors

3. Having worked at an art gallery for so long, Carl has become friendly with the gallery's bankers, who like him. Even more importantly, they support his business idea and agree to lend him money.

    A. financial institution

    B. credit card

    C. SBA-backed loan

    D. public stock offering

Correct Answers: 1. A; 2. C; 3. A

After his brainstorming session, Carl decides to finance his business in the following ways:

- He will ask friends and family for small loans to help him stay afloat as he seeks other funding. He will promise to pay back these loans with interest within the next two years.
- He will start a GoFundMe page to raise money in exchange for a free piece of small art when the site is launched.
- He will visit his local bank and attempt to secure a $25,000 loan to cover expenses for the next six months. He will attempt to secure backing for his loan from the SBIC program.
- He will get in touch with some people (angel investors) his Aunt Sam has told him may be interested in funding his venture.

Resources for a Start-Up

Carl continues talking to Tina about financing his business.

All along the way in planning, entrepreneurs can find forms of support from incubators, accelerators, and enterprise zones, and strategic partner investments.

Incubator

An *incubator* is a facility that offers small businesses low-cost offices with basic services, such as secretarial and accounting services and legal advice.

Some incubators are sponsored by state and local economic development departments. Incubators help small businesses by providing them with the support

It's amazing how quickly expenses add up when you're starting a business.

To manage your expenses, you have to take advantage of every break you can get. Let's talk about some other resources available to small businesses.

they may need when they are first getting started, often at lower prices than in the general market.

The value of incubators: When starting a new business, the option to use business incubators can be cost-effective and increase visibility with potential partners, customers, and employees.

Accelerator

Much like an incubator, *accelerators* are programs that provide entrepreneurs with a range of support that may include funding, office space, mentors, and resources to scale operations.

Accelerators provide many services, support, and funding opportunities for start-ups. Their model is to identify new companies, enlist them as protégés, and add training, contacts, and strategic partners that provide a runway toward launch and early growth stages.

Essential services such as internet access, office space, leadership training, and funding opportunities are the typical range of help entrepreneurs can leverage to advance their businesses.

Enterprise Zone

An **enterprise zone** is a specific geographic area in which government tries to attract business investment by offering lower taxes and other government support. Some are known as *urban enterprise zones* because they are located in economically distressed areas in the industrial or commercial parts of cities.

These policies have a mutual benefit for neighborhoods in need of growth, expansion, or revitalization and assists with social well-being, through new jobs and infrastructure improvements while aiding new companies.

Strategic Partnership

A *strategic partnership* occurs when new partnerships unfold, and a business can capitalize on the relationship with cash infusions if their product or service brings great value to the partner.

For example, when a complementary product needs to be integrated or upgraded, the partner will assist with these costs. Both will benefit from increased sales and entrance into a new market or become more competitive through the combination of their relationship.

Decisions

1. Suppose that Carl decides to locate his business in a certain town in Colorado, where business owners can get an investment tax credit, a job training tax credit, a new business facility tax credit, a vacant building rehabilitation tax credit, and many other benefits. That Colorado town is considered _____.

   A. a maquiladora

   B. an enterprise zone

   C. an incubator

   D. a Small Business Sanctuary

   Correct Answer: 1. B

Lesson 1-4

Understanding Small Business Strategy and Operations

By the end of this lesson, you will be able to:

- Explain the elements that define a small business.
- Summarize how entrepreneurship relates to small business activities.
- Describe the key benefits of starting a home-based business.
- Summarize the considerations when purchasing an existing business.

Owning a Small Business

"How Do I Go from a Start-Up to a Small Business?"

Carl has successfully launched ARTistic.com. His business is now evolving from a start-up to into a small business. Carl is excited to reconnect with his Aunt Samantha, a successful technology executive, to talk about this exciting phase of his business.

You've got an exciting challenge underway!

Now that you've got the business funded and launched, your job is to ensure its sustainability through strategic growth and effective operations.

I am proud of you, and I can see you have the talent, commitment, and passion.

Thanks, I am really driven to succeed. I know I can make it work because I'm offering a unique service of putting art lovers in direct contact with artists from around the world.

Yes, and as you continue to gain traction and build success I wouldn't be surprised if you are approached by individuals who would be interested in buying your business.

> Really?! You think that someone may want to buy ARTistic from me!

> Yes, for some entrepreneurs, purchasing a business and growing it is more aligned with their strengths than engaging in the start-up process.

In this lesson, we will learn about small businesses and the different opportunities individuals have related to starting, growing, or acquiring a small business.

The World of Small Business

The key parallel between small business owners and entrepreneurs is that they are driven problem-solvers that are motivated to provide products or services that are quantitative and qualitative by design.

In the early stages of an entrepreneurial venture, the entrepreneur assumes more risk by chasing rapid growth and high returns with innovative solutions.

Their ideas and unique perspectives on bringing new solutions to the market typically focuses on improving business operations through lower operational costs and increasing sales in addition to considering and supporting the needs of society.

As the business becomes more established, the entrepreneur spends more time focused on strategy, execution, and operations around growing the existing company and the established products or services.

Ultimately, the entrepreneur has direct impact on the decision-making process across the operations.

What Constitutes a Small Business?

According to the Small Business Administration (SBA), the U.S. government defines a **small business** as a business that:

- is independently owned and operated
- is not dominant in its field of operation
- meets certain criteria set by the SBA for number of employees and annual sales revenue.

A small business almost always has 100 or fewer employees, but according to some definitions, a small business may have as many as 500 employees.

Like Carl, many people run their small business out of their own homes. In fact, about 53% of new small businesses begin in the home, are financed on less than $10,000, and start with leased or used equipment. Additionally many of the business operations applications such as accounting, payroll, customer payment systems, and so on are available as pay-as-you-use services. However, some businesses require special offices and facilities to run. This is because small businesses are found in every sector of the U.S. economy: farming, manufacturing, construction, wholesaling, retailing, services, and high technology, among others.

The Role of Small Business in the U.S. Economy

"Small business is the backbone of America," the saying goes. But is it true? According to the SBA, the answer is yes.

Small businesses play a powerful role in the U.S. economy. In reality, small businesses represent the majority of U.S. sales. Small businesses also are a major source of jobs to communities across the country. In addition, small businesses help counteract the impact of downsizing by larger companies. Finally, small businesses drive growth and innovation in the economy.

Operating a Home-Based Business

The advancement in technology and cloud-based applications along with the impacts of the coronavirus pandemic have created more opportunities for individuals to pursue the creation of home-based businesses.

"National Federation of Independent Business, "Small Business Facts."

Considerations for Starting a Home-Based Business

Many businesses can be started at home. According to the SBA, the average start-up cost is $30,000, but many micro-businesses can get started for as little as $3,000 or less.

Many home-based businesses are one- or two-person operations, such as online bloggers and podcasts, online stores, and a variety of professional services ranging from accountants and consultants to software developers and interior designers.

Essentially, if an individual has a skill that can be deployed from home and is willing to invest in the right tools and technology, they can most likely create a home-based business in today's economy.

Benefits of Starting a Home-Based Business

Owning a home-based business, like any entrepreneurial endeavor, allows individuals to actually dive into their interest and passion. With diligence, hard work, proper planning, and execution, the rewards can be incredible and provide a great sense of fulfillment.

- Additionally, because startup costs are low, you can examine the opportunity before making a large investment.
- Another benefit is you can begin working on your business part-time while you maintain your current job.
- Along with being in charge of the business, the responsibility for goal setting and achievement gives entrepreneurs flexibility and rewards. The ability to define your own schedule is valuable, especially for creating your preferred work–life balance. By eliminating commute time, a home business owner can work during any time slot each day or night, you have complete control over your daily schedule.
- With a home-based business, the typical costs for physical office space, insurance, utilities, furniture, etc. are reduced or eliminated completely. These costs are also deductible from home expenses when tax filing is required.

The Challenges of Working from Home

Working at home has certain challenges, such as trying to manage your time and keep your work life and personal/family life balanced. Some suggested rules:

- **Dress for work:** Some people do just fine dressed in their pajamas, but others find that taking a shower and putting on office clothes gives them the discipline they need.
- **Set a routine and stick to it:** You should get to your desk at about the same time every day, check your email, contact your clients, and perform your work just as you would in an office in order to achieve daily goals and objectives that grow the company.
- **Connecting with others**: Email, video conferencing, and phone calls create an opportunity to engage with others. It is important for individuals not to spend too much time isolated from connecting in person. Therefore, home-based business owners may want to schedule some in-person meetings, client visits, lunches, shared work space, or other meetups.

Decisions

1. Carl will be able to work from home because his business, ARTistic, is _____.

    A. a brick-and-mortar retail store

    B. a nonprofit organization

    C. both domestic and multinational

    D. a web-based store that connects artists directly with buyers

Correct Answer: 1. D

Buying a Small Business

Individuals interested in owning a business but not necessarily starting something from scratch can consider purchasing an existing business.

There are two reasons to purchase an existing business:

- **To reduce uncertainty:** When taking over a successful business, the previous owner has already dealt with and overcome a number of the risks over time.
- **To generate profits more quickly:** When stepping into an existing well-run enterprise that has cash flow from their offerings, the new owner should be able to earn profits fairly quickly.

Tips for Taking Over a Successful Business

When taking over a successful business, the new owner will want to control expenses in order to maximize profit. Here are some inexpensive options to consider:

- Spend time with customers and employees to learn about the company, culture, and opportunities for improvement.
- Adopt a learner's mindset to understand the operations, processes, and systems of the business.
- Engage employees and leaders in creating a future vision for the company.

Here are some questions an individual should ask prior to purchasing a business.

How Do the Business and Industry Work?

You'll need to understand the business and the industry, including the products, customers, pricing, employees, competition, and the cost of doing business.

Why Is the Owner Selling, and What Do They Want to Do Next?

You'll need to find out why the owner is willing to sell. You'll also want to learn what the owner's future plans are. (Is the seller planning to start a competing business?)

Is the Price Right?

The owner may have high expectations about what the business is worth. You'll want to look carefully at the financial statements, probably with the help of a financial consultant, to determine a fair and reasonable price for the business.

Decisions

1. What is one of the reasons that Carl has chosen to start his business from scratch rather than buy an existing business?

    A. Carl wants to reduce his uncertainty and generate profits quickly.

    B. Carl wants to work from home.

    C. No businesses currently exist that offer the services that ARTistic will offer.

    D. Carl is unwilling to work more than 40 hours per week.

    Correct Answer: 1. C

Opportunities to Grow a Business

By the end of this lesson, you will be able to:

- Summarize the actions a business owner can take to increase the probability of success.
- Describe the pillars for successful business growth.

Growing Pains at ARTistic

"How Can I Help My Business Expand?"

As business has taken off, new artists have been contacting Carl to request a sales portal on his website. Carl thinks there is little risk in taking on as many artists as possible, and he feels gratified that artists think his website is a useful way of bringing their art to a wider audience.

Carl is discussing the situation with his SBA mentor, Tina, at their monthly meeting.

In the last two weeks, more than two dozen artists have contacted me.

That's fantastic. Word is getting out. Clearly, artists are networking and saying good things about ARTistic's ability to sell their art.

I have a few concerns, though. I don't have any written guidelines for what ARTistic will sell. Our value proposition is that all the art must be *affordable*, but some of these artists' work just looks ... well, I hate to say it, but cheap. I don't want ARTistic to get a reputation for selling junk.

Think of the art supplied by artists like any product offered at a retail store.

You have to make sure the products meet the company's standards and guidelines.

The other challenge you are going to face is your site is a marketplace, so you'll need to be mindful of how you regulate the market.

It also means more legal paperwork to manage because each artist has to sign a contract agreeing to ARTistic's terms.

So, this is a great opportunity, but you'll want to proceed by ensuring all your agreements are in place before sales begin and the tracking and oversight of deliverables is constant.

Its a lot to do but it will be critical for your success, you can do it!

Carl's dream is coming true—the small business that he started, ARTistic, is starting to grow. But with growth come business challenges.

Careful, controlled growth is often the best strategy because taking matters one step at a time helps business owners think, plan, and manage their expenses, all while asking the questions, "Is this growth in line with my business model and revenue plans?"

In this lesson, we will describe two key concerns of entrepreneurs: the reasons why small businesses fail, as well as the steps that entrepreneurs can take to navigate around challenges and put methods in place that can assist with stability, growth, and increase probabilities for success.

How Can Start-Ups Increase Probabilities for Success?

Small businesses are more apt to fail than big businesses are, often because small businesses are taking more risks—for example, selling products that are new and untried. But there are also at least five other common reasons for failure.

Inadequate Management Skills

Some entrepreneurs undertake a new business (opening a bed and breakfast, for example) without any actual experience in that line of work. Or they may be experienced at one aspect of a business (research and development, for instance, or sales and marketing) but know nothing about planning and cost control. Or, especially important, they don't know how to manage their cash *flow*—that is, the incoming and outgoing money.

Example: Carl needs to be careful that he knows how to manage the growth of his business or that he hires someone who does.

Lack of Financial Support

Often new ventures are seriously underfunded and don't have the deep pockets to ride out an economic downturn or other money problems. In addition, small businesses face two other financial difficulties:

- **Lack of credit:** Traditionally, banks have been reluctant to lend money to new firms because the new firms have no track record of success. For this reason, many start-up businesses are financed through personal credit cards.

- **Unfavorable economies of scale:** The term *economies of scale* refers to the savings realized from buying materials or manufacturing products in large quantities. Big businesses can buy materials in bulk, thus reducing the cost of producing each product they make. Small businesses don't have the financial clout to do the same thing.

Example: Carl needs to be careful that he seeks not just short-term funding for his business but long-term funding as well.

Difficulty Hiring and Keeping Good Employees

Small businesses often don't have the money to attract and keep good employees. Employees may leave because the firm is so small that there are no opportunities for growth and promotion. Some firms cope with this problem by offering employees partial ownership, as a restaurant owner might to a valued chef.

Example: Carl needs to realize that because of his limited budget, he may need to hire younger, less-experienced employees, at least initially, should he need additional help.

Aggressive Competition

An entrepreneur who takes a chance on a new business and then experiences success may be rewarded not only with profits but also with competitors entering the market who diminish those profits. Or the entrepreneur may have unknowingly or unwisely entered a market that is already full of cutthroat competition.

Example: Carl needs to continually scan the environment to determine whether any new competition has entered it. He also needs to be sure that he continually creates unique value for his customers so that he doesn't lose them to competitors.

Government Paperwork

Big businesses can afford to hire specialists to deal with government permits and keep government-mandated records concerning payroll deductions, taxes, expenses, safety requirements, compliance with environmental regulations, and other matters. Small businesspeople can find themselves drowning in such paperwork, which costs them time and money, takes their attention away from running their business, and carries the risk of penalties and fines for missing important deadlines.

Example: Carl needs to be sure he is on top of filing the appropriate business forms and taxes with the necessary authorities.

Tips for Keeping a Small Business Healthy

Staying in business means staying on your toes. Here are some suggestions for keeping a small business healthy.

1. **Keep good records and know when to ask for help.**

   As the business launches, it is a good idea to hire or contract with an accountant, bookkeeper, or accounting software system to help set up and record sales, expenses, payroll, inventory control, taxes, profits, and other financial matters. Accounting software packages, such as QuickBooks, can be invaluable.

2. **Stay in tune with your customers.**

   Serving customers must be a key reason for existence. As a small business, a key advantage is the ability to know customers better than big businesses do, and to quickly adapt as customers' needs change. The best way to keep customers happy is to sell a quality product or service at a value-added price.

3. **Learn how to lead employees.**

   Owners may pay employees less and offer fewer benefits compared to big businesses, but small businesses do have an essential advantage: They can keep employees engaged, motivated, and satisfied by treating them with respect, listening to their ideas, and making their work meaningful, interesting and challenging.

Decisions

In running ARTistic, Carl meets all the following challenges that may affect the success of his business. Which challenge does Carl face in each situation?

1. Carl must keep careful, detailed records of the monies he pays to each artist, because he will have to send each one a 1099 tax form each year.

    A. government paperwork

    B. lack of financial support

    C. inadequate management skills

    D. aggressive competition

2. Other websites that sell original art take notice of ARTistic and its value proposition, and they find a way to offer and advertise moderately priced art for the average-income consumer.

    A. difficulty hiring good employees

    B. lack of financial support

    C. inadequate management skills

    D. aggressive competition

3. Carl is confident in some aspects of the business he has started, including his ability to work with artists and the sales aspect of selling art to discerning customers. But he has never managed so many people (artists) before.

    A. government paperwork

    B. lack of financial support

    C. inadequate management skills

    D. aggressive competition

Correct Answers: 1. A; 2. A; 3. C

Pillars for Successful Growth

All companies should strive to align with the marketplace. The business must constantly monitor impacts on the business environment—legislation, regulation, competition, technology shifts, economic movements, social changes, and so on.

These best practice pillars should be considered as core operational elements to build a core business structure to help with these impacts and continue growth.

- Caring for Your Most Valuable Assets
- Risk and Response
- Research and Development
- Social Responsibility and Community Support

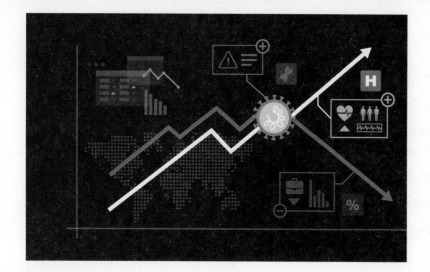

Caring for Your Most Valuable Assets

One pillar is ensuring that employees understand the experience and feel valued and that the company provides a place for them to grow, thrive, and be innovative.

This is demonstrated by having full transparency of company capital and showing the allocation available for their benefits, bonuses, education, and so on.

Employees make the company valuable because they are the ones that create unique solutions, develop ideas, extend the culture, and use all the technology and resources provided to them.

Research and Development

Research and development is another component of accomplishing operational success. These activities focus on improving existing products and services that research and user testing can inform.

The function within a company can drive innovation of new products or services and help with competitiveness in the marketplace. This activity can apply to internal processes as well.

When creative improvements occur, a company can experience lower operations costs and improved employee satisfaction, especially if the new ideas remove unnecessary steps in time-consuming production or delivery functions.

Risk and Response

As a business expands and success builds, it is essential to avoid any false sense of security. With the addition of more customers comes increased satisfaction and product performance issues.

The broader audience may have unmet expectations or product liability concerns. Competitor challenges may also surface from companies with similar products or intellectual property or legal disputes over product names, and so on, that could overlap your offerings or market space.

These are just a few examples of risks that could consume capital. Therefore, setting aside funds to address these risks is always needed. The risk pool funds can be used for legal support, insurance policies, product fixes, and so on.

Of particular note are the compliance costs for any regulatory governed products. They can change overnight and create unexpected expenses to become compliant rapidly.

Social Responsibility and Community Support

By adding the ethically justifiable factor of social responsibility as part of the mission an organization can focus to contributing to societal goals, whatever the interest the company may have because it can help build a lasting culture for the employees.

Company-based mission-oriented activities can provide more than just employment opportunities for employees, build purpose within the team members and assist the community.

Leadership should look to always create a sense of purpose, it exceeds their ideas, vision, and deliverables. Start-ups must always be driven by purpose.

Forms of Ownership

By the end of this lesson, you will be able to:

- Discuss the differences between different forms of business ownership.
- Explain the benefits and drawbacks of owning a franchise.
- Describe the different types of business mergers.

Taxes and Personal Liability: Deciding on the Legal Organization of ARTistic

"What Is the Best Form of Legal Organization for My Business?"

As Carl is consulting with web designers and making contact with artists, he is also meeting with his attorney, Brenda, regularly. Brenda brings up several issues that Carl had not even considered.

It's essential that you choose the best type of business organization, Carl. That choice will affect your recordkeeping and your tax liability, among other things.

What are my options?

Being a sole proprietor usually means less paperwork, and it's easier at tax time. But you'll be personally responsible for all of your company's debts. If your company can't pay its bills, then your creditors can come after *you*.

Ugh. I don't want that.

Fortunately, there's another alternative. You could decide to set up ARTistic.com as a limited liability corporation, or LLC.

But I'm just one person. Can an individual person really be a corporation?

Yes, and an LLC has the benefit of protecting your personal assets.

As Carl does the fun and exciting work of building the ARTistic website and contracting with artists to sell their work, he should not lose sight of the legal foundations of his business.

In this lesson, we will describe the key forms of business organization available to entrepreneurs, along with the benefits and drawbacks of each. We will also explain another popular type of ownership, a franchise. Finally, we explain two trends that often make headlines: mergers and acquisitions, in which two or more companies join to create one mega-company. By better understanding each type of business organization, Carl will be able to determine the best type for ARTistic.

Forms of Business Ownership

Three key forms of legal business ownership are (1) **sole proprietorships,** (2) **partnerships**, and (3) **corporations**.

Sole Proprietorships

A *sole proprietorship* is a business owned and typically managed by one person. In 2014, the United States had 23 million sole proprietorships, with about 70% of businesses being organized as sole proprietorships in 2016.

Partnerships

A *partnership* is a business owned and operated by two or more persons as a voluntary legal association. There are four types of partnerships.

1. In a **general partnership**, two or more partners are responsible for the business, and they share profits, liabilities (debt), and management responsibilities.

2. A *limited partnership* has one or more general partners plus other, limited partners who contribute an investment but do not have any management responsibility or liability.

3. In a *master limited partnership (MLP)*, the partnership acts like a corporation, selling stock on a stock exchange, but it is taxed like a partnership, paying a lower rate than the corporate income tax.

4. In a *limited liability partnership (LLP)*, each partner's liability—and risk of losing personal assets—is limited to just his or her own acts and omissions and those of his or her directly reporting employees.

Corporations

Corporations can be large businesses, but most are small. As noted earlier, corporations have the key benefit of limited liability for shareholders. Each shareholder's liability is limited to the amount he or she paid for stock. Creditors cannot go after stockholders' personal assets. There are four types of corporations.

1. A *C corporation* is a state-chartered entity that pays taxes and is legally distinct from its owner; this type is favored by most big businesses.

2. An *S corporation* has no more than 100 owners (shareholders), but, like a partnership, the owners are taxed only at the personal level, not the corporate level.

3. A *limited liability company (LLC)* combines the tax benefits of a sole proprietorship or partnership—one level of tax—with the limited liability of a corporation.

4. The charter of the *B corporation*, or "benefit corporation," legally requires that the company adhere to socially beneficial practices, such as helping communities, employees, consumers, and the environment.

What Are the Benefits and Drawbacks of Each Type of Business Ownership?

Each form of business ownership has its benefits and drawbacks.

Sole Proprietorships: One Owner

Financial Risk

Unlimited financial liability; owner at risk of losing personal possessions

Benefits

You may:

1. Start up the business with ease.

2. Make your own decisions.

3. Keep your taxes relatively simple.

4. End the business with ease.

Drawbacks

You may:

1. Have unlimited financial liability.

2. Have limited financial resources and few fringe benefits.

3. Have management problems.

4. Be overstressed about time.

5. Not sell or pass along the business.

Partnerships: Two or More Owners

Financial Risk

Unlimited financial liability; any general (not limited) partner is at risk of losing personal possessions

Benefits

You may:

1. Start up the business with relative ease, using a legal agreement called *articles of partnership*.

2. Have more managerial and other expertise, because different partners may bring different skill sets to the company.

3. Keep your taxes relatively simple.

Drawbacks

You may:

1. Experience personality conflicts or other disagreements among partners, which can make for unpleasant working conditions and a stressful working atmosphere.

2. Have unlimited liability (that is, the partnership's creditors can go after the partners' personal assets).

3. Have difficulty ending or changing the partnership and not be able to pass along the business to heirs.

Corporations: Unlimited Number of Owners

Financial Risk

Little or no liability; owners can lose amount they invested in the company, but not personal possessions

Benefits

You may:

1. Have little liability. That is, your liability will be limited to the amount of your investment in the company, and creditors will not be able to hold you personally responsible for the corporation's debts.

2. Get possible tax breaks.

3. Have far more financial resources, because corporations often sell stock to raise money for the business.

4. Have access to far more managerial and other expertise, because the more money raised from shareholders, the more managerial talent you will be able to purchase in the labor market.

5. Be able to easily sell ownership shares (through the stock market) and continue the business.

Drawbacks

You may:

1. Have to deal with start-up and ongoing costs and paperwork. Each state has its own set of corporation codes.

2. Be taxed twice. First, the corporation will pay a corporate tax on its profits, and then you will pay a tax on your earnings from the corporation.

3. Have difficulty ending or changing the partnership and not be able to pass along the business to heirs.

4. Have to publicly disclose financial information.

5. Have difficulty ending the corporation. Once a corporation is established, it is somewhat difficult to terminate it because of the amount of paperwork required.

One Additional Form of Ownership

While not widely used, cooperatives provide an alternative form of ownership. A cooperative, commonly referred to as a co-op, is owned and run by members with a focus on benefiting its members.

Cooperatives: Unlimited Number of Owners

Financial Risk

Little or no liability; owners can lose amount they invested in the company, but not personal possessions

Benefits

You may:

1. Have little or no liability.

2. Be able to sell ownership shares and continue the life of the business more easily.

3. Be protected against a single shareholder's dominance.

4. Find the organization more resilient because it's less prone to take risks.

Drawbacks

You may:

1. Discover that decision making takes too long.
2. Find it difficult to get bank financing.
3. Experience squabbling among members.
4. Find that the organization never makes a profit.

Decisions

1. In determining the legal organization of ARTistic, Carl decides that he wants to combine the benefits of being a sole proprietor with the limited liability of a corporation. In other words, he wants to protect his own assets from creditors. Which type of legal organization should Carl choose for ARTistic?

   A. LLC (limited liability corporation)

   B. C Corporation

   C. Partnership

   D. B corporation

Correct Answer: 1. A

Franchises: A Special Form of Ownership

Carl is meeting with his consultant, Tina, at the SBA.

A *franchise* is an arrangement in which a business owner allows others the right to use its name and sell its goods or services within a specific geographical area.

A *franchisor* is the business owner that gives others the rights to sell its products or services. The *franchisee* is the buyer of the franchise.

Business-Format Franchise

The franchisee uses the franchisor's trade name and format, following guidelines for marketing and pricing the product.

Examples: McDonald's, Starbucks, Jiffy Lube

Product-Distribution Franchise (Distributorship)

The franchisee is given the right to sell trademarked products purchased from the franchisor.

Examples: Jeep, Mercedes, John Deere, Baskin-Robbins, Exxon, Shell

> Tina, I received an interesting call today from a possible investor.

> That's fantastic. It means that investors have taken notice of your success.

> She wants to invest in a way I didn't expect, however. She wants to start a brick-and-mortar store with the ARTistic name. She wants to display the art, so that people can see it and purchase it on the spot.

Manufacturing Franchise

The franchisee is given the right to manufacture and distribute a certain product, following a formula or using supplies purchased from the franchisor.

Examples: Coca-Cola and PepsiCo give franchisees the right to manufacture their cola beverages.

What Are the Benefits and Drawbacks of Franchises?

Many would-be business owners are attracted to the established name brands of franchises. Owning a franchise offers both benefits and drawbacks.

Benefits

You may be able to:

1. Own your own business and be your own boss (to some extent—you must follow the franchisor's rules as spelled out in the franchise contract).

2. Start your new enterprise with widespread or growing name recognition.

3. Follow someone else's proven formula for doing business, which is responsible for the brand's success to date.

4. Receive marketing support. The fees you pay to the franchisor will go toward expensive TV advertisements and endorsements that you wouldn't be able to afford on your own.

5. Receive management and financial support from the franchisor, which wants you to succeed.

Drawbacks

You may:

1. Need to come up with a large initial franchise fee and other start-up costs. For some well-known brands an individual may need anywhere from $100,000 to over $1,000,000 to invest and become a franchisee.

2. Need to share your sales with the franchisor. Most chains charge monthly fees that can cost up to 12% of gross sales.

3. Have to endure close management by the franchisor, which can terminate your franchise if you fail to uphold certain standards. If you decide to sell your franchise, the franchisor has approval rights over the person to whom you sell your franchise.

4. Possibly have to deal with shady practices by the franchisor. While most franchisors are honest, some have been accused of overcharging their franchisees for supplies and assessing fees for unnecessary training.

5. Possibly be disappointed in the payoff of your franchise. Not every franchise is a success; think about the number of Starbucks coffee shops in your local area that have opened and then closed a few months later.

If you wish to start a franchise but are low on cash, do not despair. Many franchises are available for much lower fees, sometimes under $50,000. Some of these franchises include RE/MAX (real estate), Vanguard Cleaning Systems, H&R Block, and Green Home Solutions.

Decisions

1. Carl has decided to sell franchise rights to LaTonya, who will open a bricks-and-mortar store with the name ARTistic in a high-volume mall in a major city. What type of franchise has Carl authorized LaTonya to start?

    A. product-distribution franchise

    B. manufacturing franchise

    C. business-format franchise

    D. online franchise

    Correct Answer: 1. C

Mergers and Acquisitions: Paths to Business Expansion

Carl continues his conversation with Tina, his SBA business advisor.

 Carl, ARTistic has been in business for more than a year. Are you pleased with your profits?

 Yes, I'm thrilled! ARTistic has really taken off. Now I'm starting to get large offers from other companies.

 What kinds of offers?

 Some other art websites have suggested that I merge with them, and another company has offered to buy ARTistic for a very good price.

 Have you ever thought about buying another company to help ARTistic grow?

 Yes, but I've been so busy running ARTistic, I haven't had the time to do the research.

Two ways to grow a company are through internal expansion (that is, by increasing sales and capital investment) or by external expansion (when a company merges with or buys another company or companies). Two types of external expansion are mergers and acquisitions:

- A **merger** occurs when two firms join to form a new firm.
- An **acquisition** occurs when one company buys another one. A recent high-profile acquisition occurred when Microsoft bought gaming company Activision in 2022.

There are three types of mergers, as shown below.

Horizontal Merger

Two companies merge that are in the same industry and perform the same activity.

Vertical Merger

Two companies merge that are in the same industry but each performs a different activity.

Conglomerate Merger

Two companies merge that are in different industries and each performs different activities.

Mergers and acquisitions occur for four main reasons:

1. A merger or acquisition can occur as a shortcut to growth—by acquiring rather than developing a capability.

2. Mergers and acquisitions can also occur as a means of acquiring managerial talent.

3. Sometimes the goal of a merger or acquisition is to save money by consolidating operations to reduce costs.

4. Finally, a merger or acquisition can be a way to reduce taxes.

What Is a Hostile Takeover?

A **hostile takeover** occurs when an outsider (a *corporate raider*) buys enough shares in a company to take control of it against the will of the corporation's top management and directors.

Hostile takeovers are launched in two ways:

- With a *tender offer*, an outsider seeking to take over a company directly contacts the company's shareholders and offers to buy their stock at a price that exceeds the present market price.

- In a *proxy fight*, the outsider contacts shareholders and urges them to vote for the raider's hand-picked candidates for the board of directors.

Two ways firms resist a takeover are by locating a *white knight*, a buyer for the company who is more acceptable to management, or by launching a *poison pill*, taking actions designed to make the stock less attractive to the potential buyer.

What Are Leveraged Buyouts and Employee Buyouts?

There are two borrow-and-buy strategies by which a company may finance an acquisition.

In a **leveraged buyout (LBO)**, one firm borrows money to buy another firm. The purchaser uses the assets of the company being acquired as security for the loan.

In an **employee buyout**, a firm's employees borrow money against their own assets, such as their houses or their pension funds, to purchase the firm from its present owners. The employees then become the new owners of the firm.

Decisions

1. In deciding to expand his business, Carl decides to purchase another online website that sells art. His goal is to increase his revenue while consolidating expenses, thereby increasing his profit. Carl is considering a(n) _____.

   A. vertical merger

   B. employee buyout

   C. conglomerate merger

   D. horizontal merger

   Correct Answer: 1. D

Legal and Tax Awareness for Business Owners

By the end of this lesson, you will be to:

- Describe the different types of business law.
- Differentiate among the six types of taxes paid by businesses.
- Differentiate among the three types of bankruptcy.
- Recall key aspects of antitrust legislation.

The Legal Environment: ARTistic's Legal and Tax Responsibilities

"Which Laws Affect the Day-to-Day Operations of My Business?"

Today, Carl is meeting with his attorney, Brenda, to discuss recent changes in the law that affect his business, as well as some changes he'd like to make in the contracts that he signs with the artists whose work he sells on the ARTistic website.

You mentioned that you want to change the artist contract. Have you been having misunderstandings or legal troubles with the artists?

I wouldn't call them troubles, but things have been coming up that I didn't envision when I started the company.

Can you give me an example?

Here's one. According to the contract, the artists pay for the shipping and handling.

The artists agreed to that because they build the shipping price into the asking price for their art. But I never expected that the site would attract so many international customers.

I can't ask artists to pay shipping costs to China or Australia, for example. Those costs can be quite high depending on the size and weight of the artwork.

Yes, you should definitely adjust the contract, then.

Also, I didn't expect that people would want to return art.

Customers want us to pay for the return shipping, but I don't want ARTistic to pay for return shipping, and the artists don't want to pay either.

We have to adjust the contract so that it's clear that the customers are responsible to pay for shipping on returns.

We should also talk about some recent changes in tax laws that will affect you.

Sounds good. We have been allocating a portion of our gross revenue as set aside capital for paying taxes. Ugh, taxes—everybody's least favorite topic.

As Carl learned when he was starting ARTistic, every business works not only within a competitive environment, economic system but also within a legal system. While most entrepreneurs "learn as they go," it's important for them to have a solid grounding in business law before they make their first sale.

In this lesson, we will discuss the major types of business law, explaining how business disputes are resolved. We will also examine the conditions that make a contract legally enforceable, the different types of legal property, and the different types of business taxes. We close with an examination of the consumer-protection laws in the United States.

By learning the basics of contract law and taxation, and by keeping up with changes in these laws, Carl will never be caught unaware.

Business Law: What It Is, Where It Comes From

Carl continues his conversation with Brenda, his lawyer.

What Are the Components of the Legal System?

Business law, the legal framework in which business is conducted, is composed of four major types of laws.

Statutory Law

Written laws (statutes) created by legislative bodies

Common Law

Laws made by judges ruling on cases brought before them and based on precedents, the decisions that judges made in similar past court cases, which become guides to handling new cases

Administrative Law

Rules and regulations made by administrative agencies at all levels of government

International Law

Laws that govern activities between nations

How much do you know about business laws and taxes, Carl?

Truthfully, not much. I do know that it's critical for my success to become as well-informed as possible so I don't experience any surprises.

Your business is growing, and careful planning and constant understanding of the legal infrastructure and the tax systems can help us minimize the company's risk and exposure.

The United States has three court systems: federal, state, and local. Cases heard in these courts may be *criminal cases*, which are concerned with breaking laws that regulate behavior, or *civil cases*, which are concerned with the duties and responsibilities between individuals or between citizens and the government.

In both state and federal systems, an unfavorable decision in a lower court may be appealed to a higher court.

State and federal courts have three levels.

Trial Courts

Hear criminal or civil cases not specifically assigned to other courts (for example, special courts that hear probate, taxes, bankruptcy, or international trade cases)

Appellate Courts

Review cases appealed from lower courts, considering questions of law but not questions of fact

Supreme Courts

Hear cases from appellate courts. The U.S. Supreme Court also hears cases appealed from state supreme courts

How Are Legal Disputes Resolved?

In the course of business, legal disputes are bound to occur. While legal disputes are often unpleasant, U.S. businesspeople can take comfort in the fact that the country's legal system clearly spells out each party's responsibilities.

In the United States, parties to a business dispute may resolve their differences in three ways:

- They may use the **judiciary**, the branch of government that oversees the court system. That is, they may bring a court case before a judge.
- They may use **mediation**, a process in which a neutral third party listens to both sides in a dispute, makes suggestions, and encourages them to agree on a solution without the need for a court trial.
- They may consent to **arbitration**, the process in which a neutral third party listens to both parties and makes a decision that the parties have agreed will be binding on them.

Many parties agree to mediation or arbitration because these methods of resolving a dispute are much less expensive than going to trial. In addition, a trial may lead to a huge judgment against one of the parties.

Decisions

1. Carl runs his business in Illinois. Suppose the state passes a law that all companies must charge an 8% sales tax on every product they sell. This tax was made possible by a(n) _____.

   A. international law

   B. common law

   C. appellate law

   D. statutory law

2. Suppose that a rival website, ArtLovers, takes Carl to court, claiming that ARTistic is engaging in predatory business practices. Rather than go to court, Carl may decide to consent to _____, in which he and the owner of ArtLovers agree to abide by the decision made by a neutral third party.

   A. appellation

   B. adjudication

   C. mediation

   D. arbitration

Correct Answers: 1. D; 2. D

The Basics of Business Law

Many legal disputes find their remedies in contract law, tort laws, and property laws.

What Is Contract Law?

To be legally enforceable, a contract must meet each of the six conditions.

Agreement

Did a serious, definite offer get communicated, and was it accepted?

The basic parts of a contract are offer and acceptance. Agreement means that an *offer* was seriously and definitely communicated, and was then seriously and definitely understood and *accepted*. U.S. business people in foreign countries sometimes find it difficult (because of language difficulties) to know whether they really have a contract because they can't be sure that the terms of an offer were clearly expressed, or seriously and definitely understood, by one side or the other.

Consent

Was the offer accepted voluntarily, with no fraud or duress?

A contract is legally enforceable if no fraud or duress (coercion) was used—that is, if no party was pressured, threatened, or misled during the course of negotiations. For example, if you ordered a mountain bike from a bicycle shop, you have the legal right to refuse delivery of it if the merchant tries to pretend you ordered a racing bike.

Capacity

Did the parties have the capacity, or competence, to negotiate?

Both parties must be competent—of legal age and of sound mind—for a contract to be legally enforceable. For example, a contract negotiated with a child or with someone suffering from Alzheimer's disease or under the influence of drugs is not a valid contract.

Consideration

Were the items exchanged in the contract items of value?

The term *consideration* means promising to do a desired act or refrain from doing an act you are legally entitled to do in return for something of value, such as money. An important part of a legal contract is that consideration to be involved— that is, there is an exchange of items of value. The consideration need not be money, but if the exchange involves a trade or barter, the services or goods being exchanged must have some value.

For example, if you agree to do office work in your neighbor's computer business for free (to gain a little experience), but then you back out at the last minute, your neighbor cannot sue you, because there was no consideration. However, if your neighbor agreed to trade your office work for his tutoring you in computer programming, both services represent something valuable.

Legality

Did the contract involve a legal transaction?

To be legally enforceable, a contract must be lawful. A contract involving such unlawful activities as price-fixing, stolen goods, or illegal drugs cannot be enforced.

Proper Form

Was the contract prepared and signed in a form required by the law?

While most contracts are written, some oral contracts are just as valid as written contracts. However, some agreements require a written contract, such as the sale of land and the promise to answer for the debt of another person.

What Is Breach of Contract?

In a breach of contract, when one party fails to follow the agreement, the other has the right to certain remedies, including specific performance, discharge, and damages (a monetary settlement).

What Is Tort Law?

Tort law applies to wrongful injuries in business relationships not covered by contract or criminal law.

Torts, or civil wrongful acts that result in injury to people or property, may be intentional or due to negligence:

- An ***intentional tort*** is a willful act resulting in injury.
- ***Negligence*** is an unintentional act that results in injury.

What Is Property Law?

Property, or anything of value for which a person or firm has right of ownership, falls into four specific categories, as shown in the following infographic.

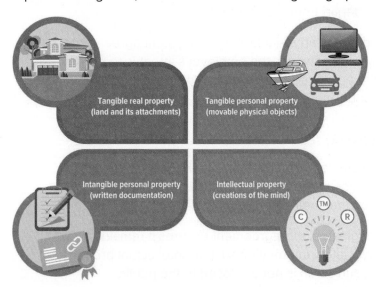

Note that there are three types of intellectual property, as the infographic below shows.

What Is the Uniform Commercial Code (UCC)?

The **Uniform Commercial Code (UCC)** is a set of U.S. laws designed to provide uniformity in sales and other commercial law and to describe the rights of buyers and sellers.

Two parts of the UCC are *warranties*, promises by a seller to stand by its products, and *negotiable instruments*, which are transferable promise-to-pay documents.

Legal Agreements Often Used in Business

The following types of contracts and agreements are typically utilized in a business environment.

Employee Non-Compete Agreement

The *employee non-compete contract* is a legally binding arrangement between an employee and their employer.

The essence of the agreement is structured in a way that precludes the employee from engaging in a competing role in a similar market, either as an independent or with a new company, once they depart from their current employer.

Frequently, these restrictions appear in an all-inclusive employment agreement as opposed to a separate document; however, the employer's intent is to build a restrictive covenant whereby the employee agrees, before becoming employed, to not to enter into or start a similar business.

Non-Disclosure Agreement (NDA)

The *non-disclosure agreement* commonly referred to as an NDA is basically an information confidentiality protection agreement that binds the employee from disclosing proprietary information that their employer owns.

Regardless of where the data originated, this agreement ensures it remains private. The agreement ensures that proprietary information such as intellectual property, business know-how, processes, and so on is not disclosed to the public.

Strategic Partnership Agreement

The *strategic partnership agreement* solidifies the relationship between two organizations, typically commercial enterprises, that defines the business relationship.

It specifically defines who the leadership is or will be, identifies the key decision-makers from each business, the definition of product offerings, licensing, management and control of intellectual property, where financing, if needed, will come from, and responsibilities for marketing, and distribution plans, and so on.

Strategic partnerships may vary in form. They may be a simple verbal agreement, deep contractual alliance, or joint ventures that take equity stakes in each other's companies.

Service Level Agreements (SLA)

Service level agreements, commonly referred to as SLAs, are client contracts that state what the customer will get, when they will receive it, and at what level of quality and cost. This commitment is a common requirement and agreement given to the customer as agreed upon in the sales contract.

The more terms covered in an SLA, the greater the comfort a client has for deliverables and support when it comes to products and services purchased from a business.

Decisions

1. Carl wants to ensure that no other business can use the brand name ARTistic. To maintain his rights to the ARTistic brand name, Carl should obtain _____.

    A. a patent

    B. a trademark

    C. a copyright

    D. a tort

2. Indicate the element of a legally enforceable contract that is described in each of the following.

    The contract that Carl's artists sign definitely involves compensation for both Carl and the artist. Carl advertises each artist's work and has made a substantial investment in a beautiful website that shows each piece of art to its best advantage. The artists do not pay to have their work displayed on the website; rather, Carl acts as a seller of that art, and he deducts commission only after a piece of art has sold.

    A. agreement

    B. consideration

    C. proper form

    D. legality

3. Indicate the element of a legally enforceable contract that is described in each of the following.

    At ARTistic, each artist understands that the contract is an agreement of terms. The artists receive a written contract spelling out all the terms in detail. Both Carl and the artist sign the contract, indicating their acceptance of the offer and the terms.

    A. agreement

    B. proper form

    C. consideration

    D. legality

4. Indicate the element of a legally enforceable contract that is described in each of the following.

   All of the artists who sign Carl's contract are adults of sound mind who have the right to negotiate with Carl regarding its terms. Any artist who is not satisfied with the contract terms is not obligated to sign the contract or sell art through ARTistic.

   A. proper form

   B. capacity

   C. consent

   D. agreement

5. Nobody forces artists to sign Carl's contract or to sell their art through ARTistic. Artists sign the contract of their own free will.

   A. Agreement

   B. Consent

   C. Proper form

   D. Legality

Taxes, Bankruptcy, and Consumer Protection

Carl continues his conversation with his lawyer, Brenda.

I want to make sure that I have a good understanding of the role tax laws and regulations play in our business. It is important we follow the law.

Yes, taxes can be a heavy burden on any business. But keep in mind that they pay for government services, like highways and police protection.

Fortunately, my accountant makes sure all the taxes are paid on time from our tax allocation funds that we set aside from gross revenues.

Business people must also be aware of laws regarding taxes, bankruptcy, and consumer protection.

What Are Taxes?

Taxes are used to pay for public services.

Income Tax

Income taxes are taxes paid on earnings received by individuals and businesses. They constitute the largest source of revenue to the U.S. federal government.

Property Tax

Property taxes are taxes paid on real estate owned by individuals and businesses, as well as on certain kinds of personal property.

Sales Tax

Sales taxes are collected by retail merchants on the merchandise they sell. These taxes are usually set as a percentage of the item's price.

State governments raise most of their revenue from sales and income taxes, though state sales taxes vary widely. Montana and New Hampshire have no state sales tax, while California has the highest state sales tax at 7.25%.

Excise Tax

Excise taxes are based on the value of services or property other than real estate, such as airline tickets, gasoline, and firearms. When applied to beer, liquor, and cigarettes, these excise taxes are sometimes called *sin taxes*. When applied to yachts, expensive cars, and fur coats, these taxes are sometimes called *luxury taxes*.

Value-Added Tax (VAT)

Also known as a *goods and services tax*, a *value-added tax* is a type of national sales tax—a consumption tax that is levied at each stage of production based on the "value added" to the product at that stage. The VAT has long been used in Europe.

What Is Bankruptcy?

Bankruptcy, the legal means of relief for debtors unable to pay their debt, may be voluntary, which means that debtors may obtain relief from creditors; or involuntary, which means that creditors may seek to have debtors declared bankrupt because of the debtors' inability to pay.

There are three types of bankruptcy, as summarized below.

Chapter 7 Bankruptcy

Forces debtors to turn over assets to a trustee for division among creditors

Chapter 11 Bankruptcy

Allows companies to reorganize and continue business operations while paying creditors a portion of their debts

Chapter 13 Bankruptcy

Allows individuals and small businesses with debts below a certain amount to repay their creditors over a period of time under a court-approved plan

What Are Consumer-Protection Regulations?

Consumer-protection laws are concerned with protecting buyers' rights. They cover food and drugs, labeling and packaging, consumer product safety, pricing (including price fixing and predatory pricing), and credit.

Antitrust law, which is designed to keep markets competitive by deterring big businesses from driving out small competitors, is based on four important pieces of legislation, as summarized in the following table.

Antitrust Legislation

Decisions

1. Carl must pay a tax on every piece of art sold through the ARTistic website. In other words, customers are paying a(n) _____ tax.

   A. property

   B. sales

   C. VAT

   D. Income

2. Carl must pay taxes to the government based on how much money he earns in a year. In other words, Carl is paying a(n) _____ tax.

   A. VAT

   B. sales

   C. excise

   D. income

Correct Answers: 1. B; 2. D

Entrepreneurship: Going from an Idea to a Business: Test

1. Melissa Aboud was abruptly laid off from her dream job after less than a year. Determined to follow her heart and concerned about putting food on the table for herself and her three children, she started Sweet Dreams, an online wholesale candy company. During this transition, working hard to get her company up and running, Aboud often finds herself making decisions based on unclear or incomplete information, trying things she had never done before, and exposing herself to financial debt for the sake of her new business venture. Aboud is displaying which of the four characteristics of the entrepreneur?

   A. high empathy levels with business associates

   B. high self-confidence and belief in personal control

   C. high need for achievement and action

   D. high tolerance for ambiguity and risk

   E. high energy level

2. Saul Mousavi worked long hours to rise to the level of middle manager at Inkspot, a print vendor servicing the needs of big local clients, including advertising agencies and colleges. But now that he has started his own print shop, Papyrus Group, he has decided to take it easy and let the business run itself. Mousavi is an example of someone who lacks or needs to strengthen which psychological characteristics of the entrepreneur?

   A. low self confidence

   B. high need for personal control

   C. high need for achievement

   D. high belief in himself

   E. high energy level

3. Which activity would be used in quantitative research and evaluation?

   A. using mathematical analytics to determine the size of a target market

   B. conducting a survey of mall visitors about interest in new store offerings

   C. collecting sentiment analysis from a company's Facebook page

   D. holding a focus group to determine consumers' feelings about a product

   E. gathering feedback on an advertising campaign from selected viewers

4. Selena Kim has often dreamed of starting her own web-based business selling vintage Asian jewelry. But she realizes she must do more than just dream. Before Kim invests her money, she needs to think carefully about her business, including potential competition, who her customers will be, and whether she has the potential to make a profit. Kim needs to _____, which will help guide her thinking about this important step.

   A. tour similar, existing businesses

   B. write a business plan

   C. take a seminar in starting a small business

   D. brainstorm options

   E. control her expenses

5. "What kind of industry am I entering, and how is my idea different?" is one of the questions you should ask yourself when

   A. planning an employee buyout.

   B. coming up with a business idea.

   C. trying to get a business loan.

   D. researching properties to buy.

   E. preparing to write a business plan.

6. Asking yourself "Who are my customers and how will I market to them?" is one of the questions to ask in preparing a(n)

   A. business plan.

   B. profit and loss statement.

   C. vision statement.

   D. mission statement.

   E. advertisement.

7. A(n) _____ outlines a proposed firm's goals and the methods for achieving them, while a(n) _____ communicates to investors the story and opportunity of the business in a short, meaningful way.

   A. business plan; pitch deck

   B. pitch deck; business plan

   C. business plan; executive summary

   D. executive summary; pitch deck

   E. business plan; business model

8. Jalessa wants to start a new business selling her handmade crafts and is looking for retail space for her new venture. A location has become available for her to rent, but the current landlord needs her to sign a lease in 10 days or else he will offer it to another client. Since she doesn't have time to properly prepare a full business plan to obtain funding, what should Jalessa do?

   A. Sign the lease and hope she can get funding later.

   B. Sign the lease with a non-compete agreement.

   C. Obtain a service level agreement with the landlord.

   D. Offer a strategic partnership agreement to potential investors.

   E. Prepare and present a pitch deck to get funding.

9. In a pitch deck, the _____ slide shows how you plan on making money and provides a schedule of when you expect money to come in.

   A. traction

   B. solution

   C. business model

   D. product

   E. investing

10. Individuals, usually wealthy, who invest their own money in a startup company are known as

    A. venture investors.

    B. mom-and-pop lenders.

    C. predatory lenders.

    D. angel investors.

    E. barterers.

11. Many entrepreneurs apply for loans of $500 to $35,000 from organizations such as the Women's Initiative for Self-Employment. These loans are known as

    A. microloans.

    B. incubators.

    C. public stock offerings.

    D. personal assistance.

    E. barter arrangements.

12. La Cocina, a nonprofit organization, mentors low-income and immigrant women entrepreneurs. It helps them launch and develop their food-related businesses by offering them kitchen space and technical assistance. La Cocina is an example of a(n)

    A. venture capitalist.

    B. enterprise zone.

    C. angel investor.

    D. incubator.

    E. strategic partnership.

13. In what way are incubators helpful to entrepreneurs?

   A. They attract business investment by offering lower taxes.

   B. They offer small businesses low-cost workspaces with basic services.

   C. They help specifically identified minorities launch new businesses.

   D. They invest in new enterprises in return for part ownership.

   E. They offer short-term credit to pay off supplies or services.

14. In some geographic areas, an entrepreneur who locates in a(n) _____ can receive lower taxes and other government support.

   A. prosperous neighborhood

   B. low-rent office space

   C. incubator

   D. enterprise zone

   E. accelerated region

15. A special type of enterprise zone, known as a(n) _____ enterprise zone, is located in economically distressed areas in the industrial or commercial areas of cities.

   A. micro

   B. strategic

   C. urban

   D. incubator

   E. venture

16. While a(n) _____ offers small businesses low-cost offices with basic services, a(n) _____ takes it one step further by offering support and funding opportunities and enlisting the business as a protégé, providing training, and strategic partners to help the business launch its growth.

   A. accelerator; incubator

   B. enterprise zone; strategic partnership

   C. strategic partnership; enterprise zone

   D. incubator; accelerator

   E. angel investor; venture capitalist

17. Which form of business is a relationship between individuals or organizations typically formed by an agreement or contract?

   A. accelerator

   B. enterprise zone

   C. strategic partnership

   D. crowdfunding

   E. incubator

18. The process in which a neutral third party listens to both sides in a dispute and makes a decision that the parties have agreed will be binding on them is known as

   A. statutory law.

   B. common law.

   C. mediation.

   D. arbitration.

   E. argumentation.

19. Ashley Cano purchased a bottle of salad dressing that broke apart when she tried to open it. She cut her hand and had to get stitches. Her buyers' rights are protected under _____ laws.

   A. food and beverage

   B. defective packaging

   C. manufacturing

   D. consumer protection

   E. factory liability

20. Which law prohibits restraint of trade and monopolies?

   A. the Sarbanes-Oxley Act

   B. the Consumer Protection Act

   C. the Robinson Patman Act

   D. the Clayton Act

   E. the Sherman Antitrust Act

What To Expect

By the end of the chapter, you will be able to:

- Analyze a well-known product's total product offer.

Chapter Topics

- **2-1** Product Strategy: Determining the Total Product Offering
- **2-2** Innovation, the Product Life Cycle, and the New Product Development Process
- **2-3** Branding and Packaging

Copyright © McGraw Hill   Zapp2Photo/Shutterstock

Product Strategy: Determining the Total Product Offering

From Family Tradition to Specialty Brand

Using Product Strategy to Grow a Business

Sofia and Gabriela Sanchez, owners of Latin Flavor Factory, sell a seasoning mix based on their Mexican grandmother Abu's authentic recipe. Inspired by the product's success, these visionary entrepreneurs want to grow their business by 1) bringing new, related Latin food products into the market and 2) growing a brand that customers associate with timeless value.

Sofia Sanchez and her sister and business partner Gabriela sit at the kitchen table sipping coffee and talking about how to build on the success of their seasoning blend, Abu Mix, by developing new products.

Our success with Abu Mix is super exciting, but let's stay focused. We want to make sure we grow this company with the right products. I think we should develop a product strategy.

I totally agree.

Before we start brainstorming new products, let's go over why our Abu Mix seasoning is so successful. What do people love about it?

That's easy. Abu's recipe is delicious. And it helps people connect with their heritage. It's like a fresh breeze from the past.

Exactly. Abu Mix offers buyers unique flavor plus a link to tradition. Our customers see value in both of those features. I'd like to bring more products to market with that same value combo.

I love it! We can build an entire brand!

How will Sofia build on the success of Abu Mix with new products that offer similar value? What new products should she develop?

In this lesson, you'll learn how to use information about what Sofia's customers value to help her decide on her total product offer. When people buy a product, they are attracted to a combination of features, from look and flavor to convenience to emotional appeal.

What Do Consumers Consider in a Total Product Offering?

All the features that potential buyers evaluate in a product are called the **total product offering**, also known as the *value package*. When you know what customers value, you can develop products that they'll want to buy. The factors of a total product offering are shown in the nearby figure.

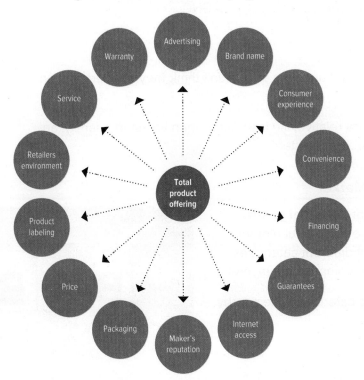

How Do Marketers Categorize Products?

There are two major markets for goods and services. The two markets are categorized as the business-to-consumer (B2C) market and business-to-business market (B2B).

The Business-to-Consumer Market

The **consumer market** consists of all those individuals or households that want goods or services for their personal use. These consumer goods and services fall into four general classes.

Convenience

Convenience: Purchased frequently and easily.

Example: Products found at a local 7/11, Walgreens, or gas station.

Shopping

Shopping: Purchased after consumers make comparisons.

Example: Products purchased online at sites like Amazon or at retailers such as Forever 21 or Anthropologie.

Specialty

Specialty: Products requiring special effort to purchase; buyers go out of their way to obtain them.

Example: Products that are unique and may only be purchased at certain locations or certain websites, such as nutritional supplements at GNC.

Unsought

Unsought: Products consumers are unaware of or didn't think they needed until an event triggers the need.

Example: Home health care for a relative, life insurance when you start a family, an emergency kit after seeing a recent disaster.

The Business-to-Business Market

The business market consists of **industrial goods**, or business goods products used to produce other products. There are a variety of different categories of B2B products and services that comprise the market.

Read more about each specific category in the nearby table.

Business Product Classifications

Category	Examples
Installations	New buildings, heavy machinery
Capital items	Industrial robots, long-haul tank trucks
Accessory equipment	Computers, copiers, desks
Raw materials	Steel, wood, oil, wool, corn
Component parts	Computer chips, batteries, switches, tires
Process materials	Industrial glue, food preservatives
Supplies	Paper, printer ribbons, cleaning agents
Business services	Legal, financial, research, online

Decisions

Sofia and Gabriela want to develop new products that will deliver value by offering the best balance of costs and benefits to the consumer.

Sofia and Gabriela brainstorm new product ideas.

So how do we come up with new products that appeal to our customers' needs and desires? Products like Abu Mix connect customers to their Latin heritage through delicious, authentic ingredients and a sense of tradition.

When customers think of Latin Flavor Factory and Abu Mix, they think of authentic Latin cooking, right? Like our grandmother Abu's special salsa, for one thing.

I loved that salsa. Didn't she use mango? And a hint of cumin? Too bad we don't have Abu's secret recipe. She always said she carried it in her heart.

Actually, I asked her to write it down for me last Christmas. Didn't I tell you?

Gabi, you're amazing. Sneaky, but amazing.

1. Given what you know about Latin Flavor Factory's proposed new product, Perfecto Salsa, which of the following are the three factors of its total product offering?

    A. Useful in an emergency

    B. Honors tradition

    C. Unique and unusual

    D. Easy to find anywhere

    E. Delicious flavor

So, thinking more about Perfecto Salsa, Sofia and Gabriela are convinced that it meets several customer needs. They feel that it will be different and taste better and that the market is ready for this new addition.

2. Which class of consumer goods do you think the Sanchez sisters should focus on as a follow-up to Abu Mix?

    A. Speciality

    B. Shopping

    C. Unsought

    D. Convenience

Sofia understands that every new product isn't a winner. To increase her chances of success, she wants to introduce more than one new product. She consults with her sister while they hit the treadmills at the gym.

So, Gabi. What else?

How about coffee? Everyone loves good strong Latin coffee.

I don't think that would work; too much competition. But some type of drink might be good. Maybe a powdered drink mix, Latino style?

Excellent! How about a *batido* mix? You know, a *batido* is like a smoothie. But we could make ours special with traditional, authentic flavorings. And it should be good for you, right? Not too much sugar or fat.

Exactly. You know, this gym is always super crowded. People really care about staying in shape, and they want healthy foods that taste awesome. Some of those protein drinks on the market just taste, well, *meh*. Why not make our new smoothie mix a nutrition supplement with a Latin flair? We could call it Batido Body Blast.

I love it! Busy people who care about their Latino roots will go out of their way to buy it, especially if it's healthy. Let's go for it.

OK, but how exactly? What type of consumer market should we target?

Correct Answers: 1. B, C, E; 2. A

Sofia and Gabriela agree that it would make sense to offer their new products wherever the current Abu Mix is offered. But they also believe that the batido mix might find a niche in fitness centers and gyms.

Innovation, the Product Life Cycle, and the New Product Development Process

By the end of this lesson, you will be able to:

- Describe the role innovation plays in today's business environment.
- Determine the stage of the product life cycle that best characterizes a product.
- Develop a new product by engaging in the six stages of product development.

So Many Products, So Little Time

How Will Latin Flavor Factory Handle Success?

Sofia Sanchez, co-owner of Latin Flavor Factory, is happy with the success of her first product, Abu Mix. Recently she and her sister and business partner Gabriela came up with two new product ideas, Perfecto Salsa and Batido Body Blast. Sofia is excited about the idea of expanding her company, but she's also a bit worried.

> We are thinking about growing. Are you sure we're doing the right thing, Gabi?

> Absolutely. After all, it's been a few years, and we still have only one product, Abu Mix. Don't you want to do more to bring authentic Latin products to market?

Of course. But what if the products fail? What if customers won't buy them?

It could happen, but it's all part of innovation. Constantly offering customers something new. Changing things up.

Won't new products distract people from Abu Mix?

I think it's going to be just the opposite. If customers see the same value in our new salsa and smoothie that they see in Abu Mix, we're golden. We plan on selling a seasoning mix, a salsa, and a smoothie. All three products honor our Mexican tradition, but they also offer people something new.

New is right. I'm not sure I've ever heard of a salsa with cumin in it. Or a nutrition smoothie that's made from a Latin recipe.

So our new products are right in line with our mission to honor Abu's Mexican heritage. Plus, they offer customers something a little different. How can we lose?

I feel better about moving forward, but it's still a little scary.

Growing means taking risks, but it's going to be worth it. You'll see.

How can Sofia and Gabriela Sanchez use their knowledge of innovation and the product cycle to guide their growth strategy as they develop and market their new products?

In this lesson, you'll help Latin Flavor Factory respond to the demands of innovation and grow as a company. Companies can't stop developing new ideas after one success, or the competition will leave them behind. Products, like people, have lives of their own. A company might simultaneously market a brand-new product, a maturing product, and a product in decline. When marketers understand the steps involved in innovation and development, they can maximize sales and stay on the cutting edge.

How and Why Do Companies Innovate?

An **innovation** is a product that customers see as being newer or better than existing products. The process of innovation takes place for three reasons:

- Product obsolescence: Existing products become obsolete as consumers' needs change or competitors add better features.

- Frequent failure of new products: Constant innovation is necessary because most new products your company introduces will not be accepted by enough consumers to be successful.

- Long development time: Some products take years to take shape.

Understanding innovation helps marketers develop the right kinds of marketing strategies. The three levels of innovation are

- **Continuous innovation**: Modest improvements, slight tweaks to an existing product to distinguish it from competitors.

- **Dynamically continuous innovation**: Marked changes to an existing product, this causes consumers to learn how to use the product or change behavior.

- **Discontinuous innovation**: The product is totally new, radically changing how people live.

Continuous innovation	Dynamically continuous innovation	Discontinuous innovation
"Slight tweaks"—generally modest improvements to an existing product	"Quite new"—marked changes to an existing product	"Brand new"—totally new product that creates major changes in the way we live

Slightly new →→→→→→→ **Radically new**

Product Life Cycle

A **product life cycle** is a model that graphs the four stages that a product or service goes through during the "life" of its marketability: (1) introduction, (2) growth, (3) maturity, and (4) decline.

Introduction

In the **introduction stage**:

- There are heavy start-up costs for production, marketing, and distribution.
- Managers concentrate on building inventory without loss of quality.
- Sales are usually low.
- There is real risk that the product will be rejected.

Growth

In the **growth stage**:

- Demand may be high.
- Managers get enough product into the distribution pipeline, maintain quality, expand sales and distribution.
- Competitors rush to get their competing products to market.

Maturity

In the **maturity stage**:

- Sales and profits start to level off due to competition.
- Managers concentrate on reducing costs and efficiency to maintain the product's profitability.
- Can extend the life of the product by tinkering with its various features.

Decline

In the **decline stage**:

- The product loses popularity.
- Managers withdraw it from the marketplace.
- Managers apply the expertise they have gained to launch other new products.

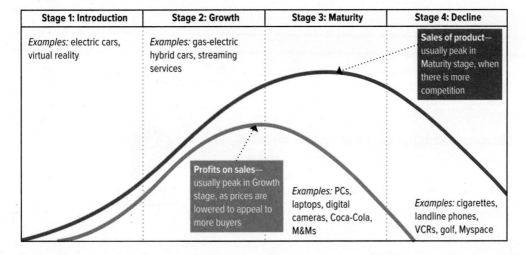

Stage 1: Introduction	Stage 2: Growth	Stage 3: Maturity	Stage 4: Decline

Examples: electric cars, virtual reality

Examples: gas-electric hybrid cars, streaming services

Sales of product— usually peak in Maturity stage, when there is more competition

Profits on sales— usually peak in Growth stage, as prices are lowered to appeal to more buyers

Examples: PCs, laptops, digital cameras, Coca-Cola, M&Ms

Examples: cigarettes, landline phones, VCRs, golf, Myspace

Decisions

1. Consider Abu Mix, the Latin Flavor Factory company's original product. The Sanchez sisters are thinking about tweaking the recipe as market research shows customers are looking for a slight more "kick" in the mix. This is an example of a _____.

   A. dynamically continuous innovation

   B. discontinuous innovation

   C. continuous innovation

Sofia approaches Gabriela, who is hard at work taking phones orders for Batido Body Blast.

Hi! I'm off the phone—for now. Batido Body Blast smoothie mix is flying off the shelves. People love the idea of a Latin Flavor Factory nutrition drink so much they can't keep enough in stock to satisfy demand. Abu would be so proud of us!

Yes, she would. But—

What? You look worried.

I just did my analysis of the past month, and sales of Abu Mix are leveling off. I was afraid this would happen!

It will be okay. Abu Mix is going through a normal part of the product life cycle.

Maybe so. But that puts more pressure on us to keep these two new products of ours rolling.

The phone rings, and Gabriela grins at it.

Batido Body Blast is hot right now, so that's a relief. And remind me to hire a customer service rep to take all these orders.

She picks up the phone.

Latin Flavor Factory, Gabriela speaking. How can I help you? Absolutely, how many cases would you like?

2. Batido Body Blast, one of the two newest Latin Flavor Factory products, is currently at the _____ stage of the product life cycle.

   A. growth

   B. maturity

   C. decline

3. Abu Mix, the company's original product, is currently at the _____ stage of the product life cycle.

   A. maturity

   B. introduction

   C. growth

Correct Answers: 1. C; 2. A; 3. A

New Product Development

A company's lifeblood is the constant creation of new products. The process consists of six steps.

| 1 Idea generation | 2 Product screening | 3 Product analysis | 4 Product development | 5 Test marketing | 6 Commercialization |

Idea Generation

During **idea generation**:

- Companies may develop new products internally by listening to employees or through research-and-development departments.

- Companies may get ideas from customers, suppliers, and competitors.

Product Screening

Product screening requires:

- Dropping product ideas that don't fit the company's product mix.
- Dropping product ideas that are too expensive, will take too long to execute, or won't generate enough sales.

Product Analysis

Product analyses involve:

- Doing cost estimates to calculate the product's possible profitability.
- Taking into account cost of materials, production expenses, impact of competitors, and potential sales.

Product Development

During **product development**:

- A product **prototype**, or preliminary version, is made.
- Some prototypes are expensive and take a long time to make.

Test Marketing

Test marketing occurs when:

- A new product is introduced in a limited form.
- Specific geographical markets are used to test consumers' reactions.

Commercialization

During **commercialization**:

- Full-scale production and marketing of the product begins.
- Organizations use information learned during test marketing. Some companies roll out the product gradually in selected geographical areas, whereas others release the product all at once to their entire market.

Decisions

Sofia and Gabriela set up a booth at a local event and offer samples of their new product prototype, Perfecto Salsa. They also ask participants to complete a brief survey. People say the unique blend of mango, cumin, and other unexpected ingredients is tangy and authentic, something they can't find anywhere else. And the product's honoring of Latin tradition is appealing, even to consumers who aren't Latino. They all agree, this product is unique and exciting.

1. In which phase of the product development phase are the sisters engaged?

   A. Idea generation

   B. Product development

   C. Product screening

2. Based on the feedback, in which of the following would you advise Gabriela and Sofia Sanchez to sell Perfecto Salsa?

   A. High-end grocery stores selling specialty goods

   B. Mexican restaurants

   C. Small shops like hair salons and thrift stores

   D. A large convenience store chain

Correct Answers: 1. B; 2. A

Lesson 2-3
Branding and Packaging

By the end of this lesson, you will be able to:

- Explain how branding and packaging help a firm differentiate its products.
- Analyze the components of a particular product's brand.
- Evaluate a brand's equity by examining the degree of brand loyalty.
- Evaluate product packaging for effectiveness.

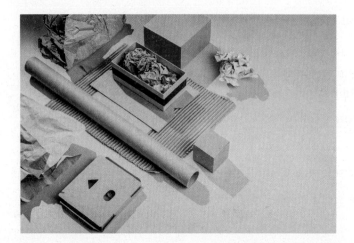

Branding a Feeling of Family

Latin Flavor Factory Launches a Family of Brands

The Sanchez sisters sit side by side on the sofa in their office, gazing up at a framed portrait of their grandmother, Abu.

I think Abu is smiling at us, Gabi.

She would be proud of us, that's for sure. Out of a family tradition, we are creating our own family.

What do you mean? We already are family.

No, I mean we're starting a family of brands. Each of our products is different, but they are all related. They all fall under the same brand, Latin Flavor Factory.

True. And back when we started thinking about adding products, you were the one who said we could start our own brand. Looks like we did it.

Yep! We can keep on coming up with new products and updating the old ones, just like Kellogg's comes up with new cereals.

But Latin Flavor Factory is so much more exciting than cereal, right?

Well, yeah. We are unique and different, and that is super exciting. But we're also about customers' attraction to the comfort and strength of tradition.

If we are going to be a real brand, we want to make sure all our products, old and new, inspire customer loyalty. Looks like we need a brand manager.

Look no further; here I am!

How will Latin Flavor Factory establish and maintain brand loyalty? Will the company's two new products generate value for the company as well as consumers? Sofia and Gabriela might need some help growing their new Latin Flavor Factory brand.

In this lesson, we'll learn all about branding and packaging. When customers decide which products to buy, they initially evaluate the brand. The brand is the name and/ or symbols that represent the producer of the product. It may be an exact name, a logo, or a representation of characters that conveys the seller to the customer.

Packaging is the combination of factors that the seller uses to wrap, box, or contain the product. Packaging may be strictly utilitarian or it may be flashy in order to excite the customer and make him/her want to buy the product. Packaging can stimulate and create imagination in the minds of the consumers in addition to merely protecting the product during shipment.

The Sanchez sisters' goal for their company is to develop a brand identity that will attract loyal customers and draw new ones in.

Creating Brands That Stand Out from the Crowd

Product differentiation involves attracting consumers by creating products that stand out as different from competitors. While companies do everything they can to differentiate their products, including pricing, in this section we will focus on branding and packaging.

- **Branding**: a unique name, symbol, or design that identifies a product
- **Packaging**: covering or wrapping that protects and promotes a product

By engaging in a multitude of engagement strategies and communication tactics a brand develops an image, personality, and reputation while providing a sense of value for its products, services.

Packaging a product provides businesses a unique opportunity to provide visual messaging about their product; as a result, companies can enhance the perceived value or communicate with their customers through messaging on the packaging.

Creating Value with Branding

Brands focus on your target audience, setting expectations and addressing value. These brands are so important that companies will spend a lot of money to protect them. Why? Four reasons:

1. **Publicize the company name and build trust:** Consumers tend to pay more for a product they have confidence in.

2. **Differentiate the company's product from competing products:** A brand is especially effective when there are a few competitors.

3. **Get repeat sales:** The more you liked the last product you bought from a company, the more likely you are to buy from it again.

4. **To make entering new markets easier:** When a company with high brand recognition introduces new products, it already has credibility.

So how do **brand managers** or product managers create unique identities for products? They use a variety of tools, including:

- **Brand names**: The parts of a brand that can be expressed in words, such as Dove and Apple.

- **Brand marks**: The parts of a brand that are expressed as graphics and symbols, like the Nike swoosh and the Under Armour UA (also known as logos).

- **Trademarks**: The parts of a brand, including names, marks, and slogans (such as, "Just do it") that are legally protected against use by others.

The Sanchez sisters have already chosen their company name, Latin Flavor Factory, as their brand name. Gabriela thinks they should also have a brand mark (or logo) and a trademarked slogan. Below are some of the brand mark and slogan ideas that Gabriela came up with to present to Sofia.

Brand Mark Ideas

- Abu's smiling portrait in a miniature form
- A simple drawing of two little girls holding hands
- A drawing of a factory
- A flying Mexican flag
- An abstract sunburst in the green and red colors of the Mexican flag

Slogan Ideas

Remember home.

Bienvenidos to flavor.

The flavor is family.

Love is the secret ingredient.

Types of Brands

As you think about the choices that Sofia and Gabriela have to make, remember that each brand is aimed at a different kind of consumer.

Branded products fall into four categories:

Brand Type	Description	Example
Manufacturer's	An individual company creates a product or service and brands the product or service using the company name.	Netflix, Uber, Sony PlayStation
Private label	A product that is sourced from a manufacturer then sold under another brand name.	Costco's Kirkland Signature brand
Family	Separate products that have differing names but are tied to the larger brand.	Kellogg's branded cereal, bars, and drinks
Individual	Products that have their own brands but may fall under a larger company.	General Motors has: Chevrolet, Buick, and GMC

Sometimes two companies may engage in **co-branding** when two noncompeting products link their brand names together for a single product. For example, Bank of America combining with Alaska Airlines to create a signature credit card. And lastly, there are *knock-off brands*—products that copy or imitate the physical appearance of other products but that do not copy the brand name or logo.

Decisions

1. The Sanchez sisters are in the process of growing their business by growing their brand. They plan on utilizing the Latin Flavor Factory in all of their branding on each of their products. Which type of branding does this best describe?

   A. Family brand

   B. Private label

   C. Co-branding

2. If the Sanchez sisters were going to have each of their products stand on its own—Abu's Mix, Perfecto Salsa, and Batido Body Blast—they would be attempting which brand strategy?

   A. Private label brand

   B. Individual brand

   C. Manufacture brand

Correct Answers: 1. A; 2. B

Measuring Brand Success

A successful brand can earn a company big profits. Marketers measure a brand's success by looking at

- **Brand equity**: The value a company gets from the factors people associate with a particular brand name, such as emotions.
- **Brand loyalty**: Consumer satisfaction with a particular brand resulting in repeat purchases.

Degrees of Brand Loyalty

Form of Loyalty	Definition	Example
Brand awareness	People recognize the product.	Although you may not know a great deal about power tools, you are more than likely aware of the brand Black and Decker.
Brand preference	People use the brand regularly.	Some travelers may prefer to fly on a certain airline over another, even if that means paying a little extra.
Brand insistence	People accept no substitutes.	A consumer purchasing a smartphone may want only an iPhone.

Highly popular brands are often **knocked off**, which occurs when individuals try to sell imitations of the product.

Decisions

As the Sanchez sisters build out their brands, they are starting to see the following comments on their Instagram account. For each statement choose the best brand loyalty description.

1. I tried your salsa the other day at Costco. It was awesome. Can you tell me where I can purchase it, out of all the Salsa's out there yours is definitely the best!

   A. Preference

   B. Awareness

2. Okay, so your company has the greatest smoothie I've ever had. I tell all my friends about it and it is the only smoothie I will every buy again, seriously EVER!

   A. Insistence

   B. Loyalty

3. I was in the store the other day looking for a mix for a new Latin dish I was making for my family and I remembered an article I read about Latin Flavor Factory. I tried and it turned out great!

   A. Loyalty

   B. Awareness

Creating Value through Packaging

Gabriela and Sofia have thought a lot about how to differentiate their products from the competition. They know that **packaging** plays an important role in branding. Overall, packaging tells your customers why your product and brand are different. In addition, packaging serves five critical functions:

Function	Why It's Important	Example
Packaging	Protects the product	Keeping the product from spoiling or leaking
	Helps consumers use the product	Packaging may have an opening for easy use, or directions on how to use the product most effectively
	Provides product information	May include nutritional information, specifications
	Indicates price and **universal product code**	Bar codes printed on the package that can be read by bar code scanners
	Promotes the product as being different from competitors	Designed package, written words, and colors to make it look superior, comparisons to competitors, features and benefits

Decisions

The Sanchez sisters are meeting with a packaging designer to go over the Perfecto Salsa packaging. Choose the following packaging goals most associated with each statement.

Because we use all organic ingredients and minimal processing, our product doesn't have a long shelf life.

I would recommend using a vacuum seal as this will maximize the products shelf life because no oxygen will get in until its open.

A. Protect the product

B. Help consumer use the product

Our market research shows people see our brand as a premium authentic product, and they use it most often at parties.

We could go with a decorative shape and a harder plastic with a large opening so it would look festive and premium and be easy to use right away.

A. Help consumers use the product

B. Provide product information

We really want to emphasize the organic, minimal processing, and local sourcing we use.

We can incorporate that into a little write up on the top of the container that says "Organic," "Locally sourced," "Finest ingredients." Also, in your nutritional facts consumers will see only a few ingredients.

A. Help consumers use the product

B. Protect the product

C. Product information

Correct Answers: A; A; C

Gabriela and Sofia have put a great deal of effort into their branding and packaging strategy, trying to ensure that they differentiate their product line. A solid product strategy will help to ensure the success of Latin Flavor Factory's products.

Products: Test

1. The term used to describe all the features that potential buyers evaluate in a product when considering whether to buy it is the

   A. basic product.

   B. product brand.

   C. value package.

   D. value components.

   E. featured elements.

2. When Thian turned 65, he decided he should purchase long-term care insurance because he didn't want his kids to have to take care of him. This is a(n) _____ product.

   A. convenience

   B. specialty

   C. retail

   D. shopping

   E. unsought

3. The _____ market consists of industrial goods, or business goods products used to produce other products.

   A. wholesale

   B. B2B

   C. retail

   D. specialty

   E. B2C

4. _____ are the large, long-lasting pieces of equipment a business purchases, like a commercial oven in a restaurant.

   A. Raw materials

   B. Capital items

   C. Accessory equipment

   D. Component parts

   E. Supplies

5. In _____ innovation, the product is totally new, radically changing how people live.

   A. improvement innovation

   B. discontinuous innovation

   C. innovation maturity

   D. radical innovation

   E. innovation decline

6. During the _____ step of the product development process, companies may develop new products internally by listening to employees or through research-and-development departments.

   A. idea generation

   B. product screening

   C. product analysis

   D. new product development

   E. test marketing

7. Broderick decided his company's new product just didn't fit the company's product mix, so he cut it from the product line. This most likely happened during the _____ stage of the product development process.

   A. idea generation

   B. product screening

   C. product analysis

   D. new product development

   E. test marketing

8. Horace invented a product that kept runners cool by inserting refrigerated pouches into their clothes. He named it Cool Runs. He wanted to get legal protection for the name, so he applied for a

   A. trademark.

   B. copyright.

   C. registered label.

   D. brand mark.

   E. legal injunction.

9. Kellogg's branded cereal, bars, and drinks fall into which category of branded products?

   A. manufacturer's

   B. private label

   C. family

   D. individual

   E. edible

10. Tonja has never even ridden a motorcycle, but she recognizes the name Harley Davidson. Tonja demonstrates brand

    A. value.

    B. equity.

    C. preference.

    D. awareness.

    E. insistence.

11. Alcides is considering the purchase of a new pickup truck. He is leaning toward buying a Ford as he appreciates Ford's warranty and service package. Ford's advertising also resonates with Alcides and he trusts the Ford name. All these features are best described as part of Ford's

    A. product cycle.

    B. value package.

    C. B2B service.

    D. brand mark.

    E. consumer market.

12. Tribiotic Medical Supply sells medical equipment to pharmacies and hospitals across the country. Tribiotic is most likely selling through the _____ market.

    A. B2C

    B. industrial

    C. B2B

    D. specialty

    E. consumer

13. NorthShore Woodworks utilizes wood and steel to build custom-crafted home furnishings. The wood and steel would fall under which class of business products?

    A. raw materials

    B. accessory equipment

    C. process materials

    D. specialty parts

    E. component parts

14. Cell phones used to be limited primarily to phone and text applications. With the advent of the smartphone, people were introduced to touch screens and countless different applications. Which form of innovation does this example best represent?

    A. discontinuous

    B. commercial

    C. dynamically continuous

    D. growth

    E. product loyalty

15. Over the past decade, more and more newspapers have downsized, consolidated, or gone out of business entirely. Newspapers are most likely in which stage of the product life cycle?

    A. maturity

    B. decline

    C. introduction

    D. discontinuous

    E. obsolescence

16. Long Life Lithium Batteries has begun producing and marketing its new "never-die" battery. Long Life Lithium Batteries is most likely in the _____ stage of the product development process.

    A. product screening

    B. commercialization

    C. test marketing

    D. growth

    E. product analysis

17. Suter Enterprises has determined that a new concept product has strong potential for profitability. How should the company proceed from here?

    A. Conduct a formal product analysis.

    B. Create a product prototype.

    C. Determine the potential market for the product.

    D. Complete a product screening.

    E. Begin full-scale production.

18. Kord needs to buy clothing for an upcoming vacation. He is usually willing to pay a little bit more in order to buy from Banana Republic. Which type of brand loyalty does this example best represent?

    A. insistence

    B. equity

    C. awareness

    D. preference

    E. maturity

19. Plateau Outerwear manufactures coats, hats, and gloves for its customers. Plateau has conducted research and knows that a large portion of its sales can be attributed to the positive feelings customers have towards the company's stance on environmentally friendly production processes. This example best describes the concept of brand

    A. loyalty.

    B. insistence.

    C. awareness.

    D. equity.

    E. preference.

20. Video game manufacturers include recommended age specifications on the packaging for games. Which function of packaging does this information best address?

    A. protection for the product manufacturer

    B. promoting brand equity

    C. provides product information

    D. helps ensure brand awareness

    E. promotes differences in competitor products

What To Expect

By the end of this lesson, you will be able to:

- Analyze how the five objectives of pricing apply to a given product.
- Calculate the break-even point for a given good or service.
- Determine which pricing strategy is being used to price a product in a business case.

Primary Pricing Objectives and Strategy

A Sweet Treat: Earning a Profit and Growing the Business

"How Can We Increase Profits and Expand Our Customer Base?"

Johann owns a small custom candy shop called Indulge. He started out creating artisan chocolates and candies with unique flavors. Recently, he was written up in a national magazine and, suddenly, he is inundated with calls from across the country.

This is a bit overwhelming for Johann, so he decides to meet with Berta, a marketing consultant, to discuss (1) how to better price his product and (2) his options for distribution.

Hi Berta, I'm really glad you were able to take this meeting.

Well, Johann, you seem to be in an exciting place—rapidly increasing sales and an expanding customer base.

Yes, but I need help. I'm not sure I know how to manage this growth.

Well, let's start by identifying your specific concerns.

First of all, when I started, I simply figured out what the candy cost me and added a markup. I'm expanding the business now and I don't think that's the way to go.

I think you're right about that, Johann, but we have several options we can review. What else?

Since the write-up, I'm getting calls from buyers across the country. Before that, I was almost strictly local. I mean, I started out selling these chocolates at farmer's markets. It was several years before I opened my small factory. I'm not sure I even know how to proceed.

Again, Johann, we can review a variety of distribution methods to see which combination of options would work best for you. Let's see how expansion might impact your pricing and distribution strategies.

Growing your business often means reevaluating your pricing and distribution strategies. The pricing of your product or service is a key element in determining the profitability of your business. However, it is not always easy to get it right. If your price is too high demand will reduce and you may price yourself out of the market. If your price is too low, your sales volume may not generate enough revenue to cover the costs associated with your business.

In addition to pricing, you need to determine how to get your product to customers. This represents the "place" aspect of the 4 Ps. For example, many local businesses rely solely on their storefront presence, or if they're business-to-business, on their local distribution path. This may not be the best option in a world that has brought us inexpensive channel development.

Johann has to measure the impact of his recent popularity in terms of pricing and distribution. Then he can assess potential profitability and expansion.

In this lesson, we will help Johann understand that pricing does more than help a business make a profit. It also can help a company to match or beat the competition, attract customers, make products affordable to certain people, and create prestige. In addition, we will help Johann learn how to use break-even analysis to determine the revenue needed to cover product costs.

What Is Pricing Supposed to Do?

Berta, my main concern is making money. So, I want to be sure whatever we do helps me make *more* money!

That makes sense, Johann. But there's more to making money than just what you charge for goods. We need to set a pricing objective and strategy.

That already sounds confusing . . . and difficult.

Nope, I guarantee that once you break this concept down into its basic components, you'll see how much sense it makes.

Pricing Objectives

No matter what type of product you sell, the price you charge your customers will have a direct effect on the success of your business. There are several *pricing objectives* that product producers—as well as retailers and wholesalers—hope to achieve in pricing products for sale.

Make a Profit and Return on Investment

Achieving a *target return on investment* is simply fancy language for making a profit, a specified yield on the investment.

For example, a business may specify that a product or service needs to make a 15% return on investment.

Match or Beat the Competition and Build Market Share

Many companies set prices simply to meet their competitors' prices.

For example, two companies selling similar products on Amazon may raise or lower their prices to keep up with each other. Keep in mind, this may have no effect on luring consumers who are brand loyal, but it will probably influence those shoppers who are price sensitive.

Attract and Build Customer Base

Also known as driving traffic—this is, of course, a principal function of pricing, which is why retailers often have sales and other pricing incentives. Low pricing can also be used to increase **market share,** the percentage of the market of total sales for a particular product or good.

For example, sometimes stores will advertise certain products as **loss leaders.** These are products priced at or below cost to attract price-sensitive customers.

Make Products More Affordable to Certain People

Low pricing can be used not only to attract customers, but also to make goods or services more widely available for those who otherwise might not be able to afford them.

For example, sporting events and movie theaters may offer special pricing to children or seniors. By making these pricing decisions, organizations can make their products more affordable and appealing to families.

Create Prestige

Higher prices help the product stand out from competitors, and this form of differentiation often makes customers value the product or brand.

For example, for some products, such as certain cars, clothes, perfumes, and watches, setting a higher price can be used to create an image of high quality or prestige.

Decisions

Berta and Johann are examining the pricing objectives for Indulge. Choose the most appropriate answer that characterizes the objective that Johann and Berta are striving to achieve in each scenario.

1. Johann has always produced mid-range priced candies. He is considering adding a line of high-end artisan chocolates, which would be marketed in the most exclusive boutiques.

   A. Make profit

   B. Make products more affordable

   C. Beat the competition

   D. Create prestige

   E. Attract customers

2. Berta suggests creating a line of lollipops with a reduced price to appeal to kids. This way, kids will be able to buy them, and as they grow up, they are attached to the brand.

   A. Create prestige

   B. Make profit

   C. Make products more affordable

   D. Beat the competition

   E. Attract customers

3. To bring in more customers, they look at starting a "buy one, get one free" promotion for holiday-themed merchandise.

   A. Create prestige

   B. Make profit

   C. Beat the competition

   D. Attract customers

   E. Make products more affordable

4. They look into offers that are better than any advertised price on boxed chocolates offered by a list of key competitors.

   A. Beat the competition

   B. Make profit

   C. Make products more affordable

   D. Create prestige

   E. Attract customers

5. They discuss making customized candy for corporate events and weddings, and for that to work, they would need to make 60% over cost for their product.

   A. Make products more affordable

   B. Beat the competition

   C. Create prestige

   D. Make profit

   E. Attract customers

Berta, I see that there's a lot more to pricing than just making money. If I move forward and expand, I'm really going to have to think about competition and attracting customers.

That's true. But right now, what's *most* important to you, besides making a profit?

Well, I think I need to know who's out there competing in this space.

I agree, Johann. Right now, you have generated a lot of interest in your products. I agree that you should take it one step at a time.

Ok, so I think that first, I need to determine if I'm really charging the right price for our candy. How could I best do that?

I think the first thing we need to do is to conduct a break-even analysis and ensure the business is making money. Then we can look at the rest of the market.

How Does Break-Even Analysis Work?

Before Johann can set a price, he has to determine how much revenue is needed to cover the total costs of developing and selling his candy. This is called a **break-even analysis**.

The purpose of this analysis is to find the **break-even point**—the point at which sales revenues equal costs; that is, the point at which there is no profit or loss.

The break-even point involves:

1. **Fixed costs**: those expenses that don't change, no matter how many products are sold. Examples might include rent and insurance.

2. **Variable costs**: those expenses that change according to the number of products produced because they are directly incorporated into the manufacture of the product or service. Examples might include cost of materials and labor.

The formula businesses use for computing their break-even point is as follows:

$$\text{Break-even point} = \frac{\text{Fixed cost}}{\text{Price of 1 unit} - \text{Variable cost of 1 unit}}$$

Remember that the break-even analysis does not provide you with a *profit*. To make a profit, you would have to increase the amount of money—a certain number of dollars to the price—above and beyond your fixed and variable costs.

Decisions

Johann spends a few hours looking at his numbers to determine his break-even point.

Fixed and Variable Costs for Indulge

Fixed costs (monthly)	$10,000 (total)
Variable costs per unit	$4.00 (average)
Selling price per unit	$20.00 (average)

Using the numbers provided, help Johann calculate his break-even point and analyze the impact of adjustments to his costs and selling price.

1. Based on this information how many items does Johann need to sell to break even?

   A. 625

   B. 476

   C. 555

2. If Johann were to reduce his variable cost by $2.00 per item, what would be his break-even quantity?

   A. 555

   B. 625

   C. 476

3. Then he ponders, what if he could keep the variable cost down by $2.00 and increase his selling price by $3.00 to $23.00. What would his break even be?

   A. 625

   B. 555

   C. 476

Now that Johann has determined his break-even point using different price points and variable costs, he can more fully consider his pricing options.

He is quickly learning that controlling costs and setting prices play a role in breaking even and making profit.

Correct Answers: 1. A; 2. A; 3. C

What Are the Primary Pricing Strategies?

Berta, I've been wondering about the price of my boxed chocolates.

What are you thinking?

Well, I've always just figured out what the chocolate cost to produce and added a few bucks to that.

As I'm sure you know, there are many additional costs—like overhead. Have you added all that in?

I think I've just estimated. But if I'm going to grow, I have to make sure I'm considering all of the pricing options.

Absolutely. Maybe we should review those?

Berta reminds Johann that choosing the correct pricing strategy will increase sales and maximize profits. She explains that there are three principal strategies for pricing a product or service. They are:

1. **Cost-based pricing**

2. **Target costing (demand-based pricing)**

3. **Competitive-based pricing**

Cost-Based Pricing

Cost-based pricing covers the costs of the product and profit. Strictly speaking, this approach to pricing ignores market forces, such as the efforts of competitors to undercut you.

Imagine a firm called Joe's Pizza Company. It costs $3.00 to make each pizza and Joe's wants a margin of 70%.

$$\text{Because: } \$3.00 \times 70\% = \$2.10,$$

The lowest price at which Joe's could sell its pizzas would be $5.10.

Target Costing (Demand-Based Pricing)

Demand-based pricing meets optimum pricing, profit, and production goals. Unlike cost pricing, target costing considers market forces. In this case, a company starts with the price it wants to charge, figures out the profit margin it wants, then determines what the costs must be to produce the product to meet the desired price and profit goals.

Joe's Pizza Company wants to sell pizza for $10. The firm needs to make a 15% profit. That means the pizza ingredients, labor, and overhead need to total less than $8.70.

$$\frac{Selling\ price}{1 + margin\ \%} \longrightarrow \frac{\$10.00}{1.15} = \$8.70$$

Competitive-Based Pricing

Competitive-based pricing helps companies compete with rivals. In this case, price is determined in relation to rivals, factoring in other considerations such as market dominance, number of competitors, and customer loyalty.

Joe's Pizza wants to pull customers in by being cheaper than all the neighborhood pizza places. The firm checks out competitors and decides to sell its pizza for $9.00 a pie. Assuming costs and overhead stay the same, Joe's will only make $0.50 profit per pie. Will it be able to stay in business?

Decisions

Choose the term that best fits the conversation between Berta and Johann.

Initially my approach was _____ to figure out what it cost to produce the candy and then add a few bucks on top of that.

A. cost-based pricing

B. target costing

C. competition-based pricing

I think when you were small and just getting going that strategy makes a lot of sense. As you scale you may want to take a look at some other approaches. For example, if you were to do _____ by starting with a price you think customers will pay for your product, then subtract out the profit you would like to make to acquire the cost you need to stay under when producing. This is a form of _____ pricing.

- **A.** target costing; cost-based
- **B.** target costing; demand-based
- **C.** competition-based pricing; demand-based
- **D.** demand-based pricing; cost-based

Yes, I started noticing when I was calculating my break-even that as my price increased or my costs decreased, my gross profit grew rapidly and I had to sell much less to break even. It's pretty awesome.

We call that the magic of margins!

If we use _____ pricing, this will allow us to increase and decrease the price based on our rivals, and there are quite a few of them in the market.

- **A.** competition-based
- **B.** demand-based
- **C.** cost-based

Correct Answers: A; B; A

Alternative Pricing Strategies

By the end of this lesson, you will be able to:

- Determine which alternative pricing strategy is being used to price a product in a business case.

Strategically Pricing to Increase Profit

"What Are Other Ways That We Can Continue to Grow?"

Johann has heard that there are strategies that a business can use to better compete in the marketplace. He's also heard there are psychological tactics that can be deployed to help persuade customers. He wants to learn even more about pricing, so he continues to meet with Berta to learn about unique pricing strategies.

Johann, I wouldn't be doing my job if I didn't explain that there are other pricing objectives and strategies you might want to consider.

Oh, there are more? Let's talk! I'm always interested in new ways to make money and gain market share.

You sure have come a long way!

Well, you've opened my eyes to all sorts of possibilities.

In this lesson, we will help Johann understand that there are alternatives to the primary pricing objectives and strategies. He will need to consider these when determining the appropriate pricing strategy for Indulge's products.

What Are the Alternative Pricing Strategies?

Pricing strategies can be developed to meet a variety of market conditions. They can be used to defend an existing market from new entrants, to increase market share within a market, or to enter a new market. So, in addition to the three principal strategies for pricing a product or service, we have seven other approaches.

Price Skimming

Price skimming tends to occur when there is little competition and a company can put a high price on the product. To recover its high research and development costs on a product, a company may resort to price skimming, setting a high price to make a large profit; it can work when there is little competition. Naturally, the big profits will quickly attract competitors.

Example: A new mobile phone.

Penetration Pricing

When a company prices the product low to attract a lot of customers and to deter competitors, it is using **penetration pricing**. This pricing strategy is designed to generate customers' interest and stimulate them to try out new products.

Example: Subscription services like Dollar Shave Club that give users reduced prices.

Discounting (High-Low Pricing)

Discounting, or high-low pricing, occurs when a company assigns regular prices to products but then resorts to frequent price-cutting strategies, such as special sales, to undercut the prices of competitors. A drawback of this strategy is that customers may tend to wait for the sales to do their shopping.

Example: Holiday sales at department stores like Macy's or JCPenney.

Everyday Low Pricing (EDLP)

Everyday low pricing (EDLP) occurs when a company prices products lower than its competitors and has no special sales. Unlike discount pricing, everyday low pricing (EDLP) is a strategy of continuously setting prices lower than those of competitors and then not doing any other price-cutting tactics such as special sales, rebates, and cents-off coupons.

Example: Everyday low prices are often seen at Walmart.

Bundling

Bundling is the practice of pricing two or more products together as a unit.

Example: A cable company that packages television, Internet, and phone service together.

Psychological Pricing (Odd-Even Pricing)

Psychological pricing is sometimes called odd-even pricing. This is the technique of pricing products or services in odd, rather than even, amounts to make products seem less expensive ($99.99 instead of $100).

Example: A sign that reads, "Only $19.99."

Credit Terms

Credit terms can also be used as part of your company's pricing strategy. You can encourage sales by offering "No payments for 12 months!"

Example: A furniture store that advertises that no payments are required for three years.

Decisions

This is really interesting, Berta. I can see all sorts of future potential here. I like the idea of _____, as I can really separate myself by setting an image with high prices and a unique product while maintaining or lowering my cost in the future.

- **A.** psychological pricing
- **B.** penetration pricing
- **C.** bundling
- **D.** price skimming
- **E.** discounting

Exactly, and if you decide to develop a new product or line aimed at a different audience you could always engage in _____ by coming in low and building awareness for the new product.

   A. psychological pricing
   B. discounting
   C. penetration pricing
   D. price skimming
   E. bundling

Yes, then as I grow my lines I can create _____ for multiple items like a box of chocolate, hard candies, and bars all together in a basket.

   A. penetration pricing
   B. discounting
   C. bundling
   D. psychological pricing
   E. price skimming

Now you are talking! When you price it, you could try something like $29.99 instead of $30.00 which would be a great way to use _____.

   A. psychological pricing
   B. bundling
   C. price skimming
   D. discounting
   E. penetration pricing

I like that, as I build my audience I can offer _____ for new customers so they can give it a try at a reduced price.

   A. psychological pricing
   B. penetration pricing
   C. bundling
   D. discounting
   E. price skimming

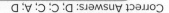

Correct Answers: D; C; A; D

The Distribution Mix: Marketing Channels

By the end of this lesson, you will be able to:

- Construct a distribution mix for a product that includes the appropriate distribution channels, intermediaries, and takes into account the four considerations of moving a product from producer to consumer.

Growing Your Customer Base through Distribution

"How Can We Expand Our Customer Base?"

Johann has a solid grasp on how he wants to price his products. Now he has to ensure the product is accessible to customers by ensuring it is distributed through the right channels. He also knows these channels may impact his margins, so he meets with Berta to discuss opportunities to optimally distribute his candy.

Johann, we haven't talked about how you reach your customers and how you get your products to your customers. What do you do now?

Well, I primarily take phone orders or sell directly here in the store. And if I have to ship, I usually use truck transport.

But as you grow and expand, you might need to have some options.

You're right. Suppose I start getting larger orders or even overseas orders, what kinds of choices do I have, Berta?

In this lesson, we will help Johann understand the different methods that he can use to get his products to his customers. Basically, Johann needs to determine which delivery methods work best for his company.

What Distribution Channels Are Available?

In marketing, you can have a great product at the right price and utilize the right promotion to get customers excited to buy what you are selling; however, if you can't deliver it at the right time, consumers can't find it at the right location, or they can't access it when they want it, then your effort will be all for naught. This is why we have the "place" category in marketing. What we are talking about is *distribution*.

There are a variety of systems by which you can convey goods or distribute to customers.

A **distribution channel**, also known as a marketing channel, is a system for transporting goods or services from producers to customers.

For example, T-shirts with slogans on them might move from producer to wholesaler to retailer to consumer or they might be marketed via the Internet directly to the consumer by the producer.

Types of Intermediaries

Within distribution channels, there are a variety of intermediaries—agents/brokers, wholesalers, and retailers. These intermediaries, or marketing **intermediaries**, are the people or firms that move products between producers and customers.

There are three main intermediaries that work with manufacturers and producers to get products to customers:

- **Agents/brokers**: Specialists who bring buyers and sellers together and help negotiate a transaction.

Copyright © McGraw Hill   kali9/E+/Getty Images

- **Wholesalers**: Middlemen who sell products (1) to other businesses for resale to ultimate customers or (2) to institutions and businesses for use in their operations.
- **Retailers**: Intermediaries who sell products to the final customer.

Ways Intermediaries Add Value to products

Type of Value (Utility Added)	What It Means
Form utility	Changing raw materials into useful products
Location utility	Making products available where convenient
Time utility	Making products available when convenient
Information utility	Providing helpful information
Ownership utility	Helping customers acquire products
Service utility	Providing helpful service

Types of Distribution Channels

All in all, we'll review six distribution channels—four for consumers and two for businesses.

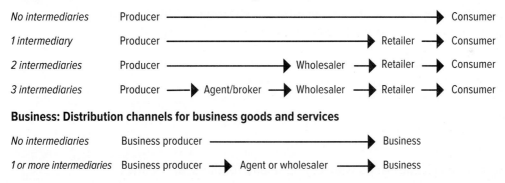

Consumer: Distribution channels for consumer goods and services

No intermediaries — Producer → Consumer

1 intermediary — Producer → Retailer → Consumer

2 intermediaries — Producer → Wholesaler → Retailer → Consumer

3 intermediaries — Producer → Agent/broker → Wholesaler → Retailer → Consumer

Business: Distribution channels for business goods and services

No intermediaries — Business producer → Business

1 or more intermediaries — Business producer → Agent or wholesaler → Business

Consumer Distribution Channels

For consumer goods, these distribution channels might include:

- No intermediaries (**direct channel**): selling directly to the consumer.
- One intermediary: Ford sells cars to dealers who sell them to consumers.
- Two intermediaries: Flower growers bring their flowers to the Flower Market, which sells to florists, which sells to consumers.
- Three intermediaries: A French winery conveys wine to a U.S. agent who conveys the wine to a distributor which sells the wine to a wine store.

Business Distribution Channels

For business goods and services, we include:

- No intermediaries: A rubber maker sells directly to a tire company, such as Goodyear.

- One or more intermediaries: A forestry company sells rolls of paper to a paper merchant who sells it to newspapers.

Determining the Appropriate Distribution Channel

So, how do you determine the right channels for your product and customers? Well, you need a **distribution strategy**, an overall plan for moving products from you to your purchaser.

The **distribution mix** is the combination of distribution channels a company uses to get its products to customers. For example, a producer might sell its products directly to the consumer or use all the intermediaries we reviewed above.

In fact some businesses have built their business around a distribution strategy. For example, Amazon has created distribution for many producers, and Costco serves as a wholesaler to its members.

Decisions

Johann and Berta are examining some different distribution strategies. Choose the distribution strategy that best applies to each statement.

1. Johann wants to be able to sell and ship to his customers from his website.

   A. Producer to retailer

   B. Direct channel

   C. Wholesaler

2. Johann is hoping to build relationships with key specialty retailers where they would order from him and they would sell in their stores to the customer.

   A. Wholesaler

   B. Direct channel

   C. Producer to retailer

3. Johann thinks that for some products it might be best to sell it to a _____ where retailers can buy in bulk and then sell it to their customers.

   A. direct channel

   B. wholesaler

   C. producer

So, Berta, why would I use an intermediary?

Intermediaries add value by saving you time and money, for instance, or by doing some tasks better than you could.

OK, I understand that. Anything else?

Well, Johann, intermediaries also add value—utility, want-satisfying ability—to products by making them more useful or accessible to consumers

Choose the utility the intermediaries would be providing to Johann's business.

4. By increasing the number of retailers, his products would be more available and convenient.

   A. Marginal utility

   B. Ownership utility

   C. Location utility

   D. Time utility

5. If Johann were to utilize a wholesaler, he could increase volume and there would be less lag time making the product available when convenient.

   A. Information utility

   B. Location utility

   C. Time utility

   D. Form utility

Correct Answers: 1. B; 2. C; 3. B; 4. C; 5. C

Forms of Distribution

By the end of this lesson, you will be able to:

- Determine which intermediary(ies) is/are most appropriate in a business case.
- Determine the appropriate retail strategy(ies) for a firm's product.

The Roles of Distributors

"Who Does What in the Supply Chain?"

Johann wants to gain a better understanding of all the potential intermediaries and learn about the value they create. He knows that the more points in the supply chain the tighter his margins will become, so he wants to speak with his consultant Berta to determine how he should leverage these intermediaries.

I'd really like to learn more about intermediaries. While they may not be pertinent for my business right now, I think they could be helpful in the near future.

Why do you say that?

Well, hopefully I'll be ramping up sales—both in terms of numbers and types of products *and* numbers of customers. And given that I want to expand my line, I see opportunities in stores and catalogs . . . well, the options are endless. But I'd sure need help. I couldn't just handle those orders from the office.

That's the way to go, Johann. The more you know, the better prepared you'll be.

In this lesson, we will help Johann learn more about intermediaries and how they might help Indulge distribute its products. Intermediaries between producers and consumers consist of wholesalers, agents and brokers, and retailers.

Role of Wholesalers

Wholesalers sell products (1) to other businesses for resale to ultimate customers or (2) to institutions and businesses for use in their operations. Retailers, by contrast, sell products directly to customers.

The three principal types of wholesalers are manufacturer-owned wholesalers, full-service merchant wholesalers, and limited-function merchant wholesalers. Let's explore each in turn.

Manufacturer-Owned Wholesalers: The Producers Do All the Selling and Distribution

A **manufacturer-owned wholesaler** is a wholesale business that is owned and operated by a product's manufacturer. Some manufacturers want to do this to maintain full control over the selling and distribution of their products or because it costs them less than using independent wholesalers.

Full-Service Merchant Wholesalers: Independents Take Over the Products and Provide All Services

A **full-service merchant wholesaler** is an independently owned firm that takes title to—that is, becomes owner of—the manufacturer's products and performs all sales and distribution, as well as provides credit and other services. Example: Beer distributors are one of these types.

Limited-Function Merchant Wholesalers: Independents Take Over the Products but Provide Only Some Services

A **limited-function merchant wholesaler** is an independently owned firm that takes title to—becomes owner of—the manufacturer's products but performs only selected services, such as storage only.

Three common types of limited-function wholesalers are (1) rack jobbers, (2) cash-and-carry wholesalers, and (3) drop shippers.

Rack Jobber: "We Provide Shelves of Products in Stores."

A **rack jobber** is a limited-function wholesaler who furnishes products and display racks or shelves in retail stores and shares profits with retailers. A candy vendor might put shelves of candy and gum in gas stations and split the profits. Magazines are often placed by rack jobbers.

Cash-and-Carry Wholesaler: "We Sell to Walk-in Customers Paying Cash."

A **cash-and-carry wholesaler** is a limited-function wholesaler that sells mainly to small retailers, who come to the wholesaler, pay cash for a product, and carry it out ("cash and carry"). Nowadays, such wholesalers take credit cards as well as cash. Staples and Costco are examples of stores that sell to small retailers.

Drop Shipper: "We Take Orders and Arrange for Shipment."

A **drop shipper** is a limited-function wholesaler who owns (has title to) the products but does not have physical custody of them. Rather, the drop shipper takes orders and has the producer ship the product directly to the customer. Sand, coal, lumber, and other bulky goods are often handled by drop shippers.

Manufacturer-Owned Wholesaler

A wholesale business that is owned and operated by a product's manufacturer. The two types of manufacturer-owned intermediaries are:

1. The manufacturer's branch office
2. The manufacturer's sales office

Example: Car dealers and tire sellers generally branch offices.

Full-Service Merchant Wholesalers

An independently owned firm that takes title to—that is, becomes owner of—the manufacturer's products and performs all sales and distribution, as well as provides credit and other services.

Example: Beverage distributors

Limited-Function Merchant Wholesalers

An independently owned firm that takes title to—becomes owner of—the manufacturer's products but performs only selected services, such as storage only. Three common types of limited-function wholesalers are:

1. Rack jobbers: Furnish products and display racks or shelves in retail stores and share profits with retailers.
2. Cash-and-carry wholesalers: Sell mainly to small retailers, who come to the wholesaler, pay cash for a product, and carry it out ("cash and carry").
3. Drop shippers: Own (have title to) the products, but do not have physical custody of them. Rather, the drop shipper takes orders and has the producer ship the product directly to the customer.

Decisions

> **Let's think this through, Johann. Which of these might be most useful to you in the future?**

> **Well, I guess it depends on how much support I need. I don't want to relinquish too much control. On the other hand, I don't want to do everything myself . . . at least, not initially.**

1. Which of the following best describes Johann's strategy?

    A. Full-service merchant wholesalers

    B. Manufacturer-owned wholesalers

    C. Limited-function merchant wholesalers

Correct Answer: 1. C

Role of Agents and Brokers

Agents and brokers are specialists who bring sellers and buyers together and help negotiate a transaction. Their value to a manufacturer or producer is their knowledge of markets and their experience in merchandising.

Agents tend to maintain long-term relationships with the people they represent. For example, sales agents (or manufacturer's agents) represent several manufacturers in one territory (but not competing products).

Brokers are usually hired on a temporary basis. Their relationship with the buyer or the seller ends once the transaction is completed. Brokers are often found in corporate real estate sales or leasing.

Agents and brokers are neither wholesalers nor retailers because they:

- Don't own—take title to—the products they handle.
- Usually don't carry inventory or provide credit.
- Earn fees or commissions (a percentage of the sales transaction) rather than profits.

Decisions

This is very interesting, Berta. I can see agents and brokers may be able to help us on sales.

Yes, in a variety of ways. Let's think of how they might help you.

1. If Johann were to hire someone on a one-year contract to help find new suppliers, which type of position would he be needing?

   A. Broker

   B. Agent

2. As Berta is trying to source a location for the business, she might need additional assistance during the next few weeks to help her find options. What type of assistance would best meet her needs?

   A. Hiring an agent

   B. Hiring a broker

Correct Answers: 1. B; 2. B

Role of Store and Nonstore Retailers

Store Retailers

The roughly 2.3 million store retailers in the United States range from those with broad product lines, such as department stores and supermarkets, to bargain retailers, such as discount stores and warehouse clubs. Store retailers also refer to specialty retailers, which can include pushcarts, kiosks, and drive-thrus.

Any of these stores might be found in shopping centers—from strip malls to super-regional centers, such as the Mall of America in Minnesota.

Types of Retail Stores

Within the broad category of store retailers, there are product-line retailers and bargain retailers. Both are explored below.

Department Stores

Department stores sell a variety of products (e.g., furniture, clothes) in departmentalized sections.

Examples: Bloomingdale's, JCPenney, Macy's, Nieman Marucs, Nordstrom, Saks Fifth Avenue

Supermarkets

Supermarkets sell mostly food and some nonfood products in specialized departments.

Examples: Albertson's Kroger, Piggly Wiggly, Publix, Safeway, Trader Joe's, Whole Foods, Winn-Dixie

Supercenters

Supercenters are giant stores that sell a wide variety of unrelated products.

Examples: Walmart and Target

Category Killers

Category killers sell a huge variety of one type of product that may drive out smaller competitors.

Examples: Best Buy, Home Depot, Lowe's, Staples

Convenience Stores

Convenience stores sell staple convenience foods and other goods at convenient locations and hours.

Example: Gas station minimart

Specialty Stores

Specialty stores sell a variety of goods in a narrow range.

Examples: AutoZone, Barnes & Noble, Bath and Body Works, Foot Locker, Gap, IKEA, Rite Aid, Victoria's Secret, Walgreens, WIlliams-Sonoma

Discount Stores

Discount stores sell a wide variety of merchandise at substantial price reductions.

Examples: Dollar Stores, Dollar Tree, grocery outlets, Target, Walmart

Warehouse Clubs

Warehouse clubs are large, warehouse-style stores that sell food and general merchandise at discount prices.

Examples: Costco, Sam's Club

Outlet Stores

Outlet stores are manufacturer-owned stores that sell discontinued or flawed products at a discount.

Examples: Burberry Outlet, Burlington Coat Factory, Liz Claiborne, Nike Outlet

Secondhand Stores

Secondhand stores sell used merchandise.

Examples: Goodwill, Salvation Army, Plato's Closet, Buffalo Exchange

Nonstore Retailers

Nonstore retailers sell merchandise in ways other than through retail stores. Even 10 years ago there were many forms of nonstore retailing, but the Internet has vastly expanded the possibilities.

Vending Machines

Coin-operated vending machines selling convenience goods, such as candy and sodas, have been around seemingly forever.

Now, of course, such push-button or lever-operated machines not only accept paper money and credit cards and make change but also sell a great many other kinds of products, such as cosmetics and electronics.

Direct Selling

Direct selling is one of the oldest forms of nonstore retailing. It consists of face-to-face selling directly to customers in their homes or where they work.

1. **Door-to-door**
2. **House-party**
3. **Multilevel marketing**

Direct Marketing

Not done face to face, but consists of selling directly to customers using mail or telephone.

1. **Direct mail marketing**
2. **Catalog marketing**
3. **Telemarketing**

Video Marketing

Video marketing is marketing to consumers on television, either through special cable TV channels or through certain programs on regular TV channels.

Consumers are given a sales pitch by a TV salesperson and urged to call a toll-free number or go to a certain website to place their order.

Online Retailing

Online retailing, also known as electronic retailing, this is nonstore retailing of products directly to consumers using the Internet.

Decisions

> Johann, I want to be sure you understand all of your options.

> Hey, the more options, the better!

> You see, there are also *nonstore* retailers that might be of interest to you.

> That almost seems like an oxymoron!

Later, Johann and Berta are considering some different selling strategies. Choose the term that best applies to the description.

1. Johann and Berta are considering the idea of having customers become sales representatives of their products in order to increase sales. Each new representative would make a commission on what he/she sells. This type of marketing falls into which category?

    A. Direct selling

    B. Vending machine representatives

    C. Direct sales representatives

    D. Direct marketing

2. If Johann and Berta decide to run sales commercials on TV and then ask customers to contact them to order the product, this would be _____.

   A. direct selling

   B. indirect selling

   C. online retailing

   D. video marketing

 I do think that—once I've established my product lines—I would like to see them carried in specialty stores.

 Why do you say that?

 Well, it's great exposure, but more than that, it's pretty much a steady sale.

 Just remember, you need to keep your line fresh and you need to work your price so that the store profits as well.

 Of course! I think I'd need to be more established before I go that route! But it's a great opportunity!

3. If Johann decides to sell his product line through specialty stores, what type of product selection would this store contain?

   A. A wide variety of special goods

   B. A narrow category of goods intended for a specific market

   C. Discounted products from a specific product line

   D. Convenience goods to the customer

4. If Johann were to try and place his goods in stores that carry a massive selection of products in a specific genre, which store type should he select?

   A. Specialty

   B. Category killer

   C. Outlet

   D. Warehouse club

5. In the future, toward the decline stage of Johann's products, he might want to sell his products through which type of retail stores?

   A. Warehouse club

   B. Specialty

   C. Outlet

   D. Department

Correct Answers: 1. A; 2. D; 3. B; 4. B; 5. C

Understanding the Supply Chain and Logistics

By the end of this lesson, you will be able to:

- Recommend the optimal method of transport and storage for a firm's product.

Figuring out the Logistics of Moving Product

"What Are Supply Chains and How Are Transportation and Warehousing Involved?"

Johann is gaining great insights on how to effectively distribute his product. He is now curious on how the product is physically distributed and how he needs to create his supply chain to grow the business.

Johann, in all of our conversations, we haven't really discussed the physical distribution of your products.

Hmmm, I really haven't had to think much about it. People either picked candy at the store, or if they called in an order, we would ship by ground.

Well, I think you know that's going to change.

Yes, I certainly *hope* so!

In this lesson, we will help Johann understand the complexity of physical distribution, or movement of all products from manufacturer to final buyer.

What Is Supply Chain Management and Logistics?

Physical distribution, the movement of all products from manufacturer to final buyer, involves a supply-chain sequence of suppliers and logistics. Logistics is the details of transportation and warehousing that make distribution happen. Transportation may be by air, road, rail, water, or pipeline, with each having its own advantages. Warehousing may be simple storage of products for long periods of time or distribution storage for short periods of time. Primarily, physical distribution begins with questions such as:

- Order processing: How quickly should orders be processed and products shipped?
- Transportation: How fast or how cheaply should goods be moved? What kind of transport—truck, train, plane, and so on—is best?
- Storage: Do we need to warehouse products at any stage? Where? Who should handle them?

Supply Chains and Logistics: Moving Products to the Final Buyers

Getting products into the hands of customers involves a **supply chain**, the sequence of suppliers that contribute to creating and delivering a product, from raw materials to production to final buyers.

With so many suppliers, kinds of transport, and materials, how do businesspeople control all this? This is the province of **supply chain management**, the strategy of planning and coordinating the movement of materials and products along the supply chain, from raw materials to final buyers.

If supply chain management is concerned with *strategy*, logistics is concerned with *tactics*—the actual movement of products. **Logistics** consists of planning and implementing the details of moving raw materials, finished goods, and related information along the supply chain from origin to points of consumption to meet customer requirements. Often this requires sophisticated computer hardware and software to determine how to do it as efficiently and as cost-effectively as possible. So, logistics involves several forms of movement of both goods and services *and* information.

1
Raw materials

2
Suppliers' plants

3
Manufacturers

4
Wholesalers

5
Retailers

6
Consumers

Inbound Logistics: Movement from Suppliers to Producers

Inbound logistics involves bringing raw materials, packaging, and other goods and services and information from suppliers to producers.

Example: Book publishers bring together the materials from authors, illustrators, and photographers to a production department for editing, typesetting, and rendering to ultimately prepare the work for the printer and binder.

Materials Handling: Movement of Goods to and from and Within Producers' Facilities

With materials handling, the movement of goods may be within a warehouse, from a warehouse to a factory floor, or from factory floor to various workstations.

Example: Book printers and binders bring together electronically prepared plates, paper, and cover materials to prepare published books.

Outbound Logistics: Movement from Producer to Consumers

Outbound logistics involves managing the movement of finished products and information from producers to business buyers and final consumers.

Example: Published books are moved from the book bindery to a warehouse and then to bookstores.

Reverse Logistics: Movement from Consumers Back to Producer

Reverse logistics involves bringing defective or unwanted products returned by consumers back to the producer or manufacturer for further handling, such as repair or recycling.

Example: Booksellers return their unsold books to publishers, who then sell them on secondary markets or destroy them.

What Are the Trade-Offs for Transportation Choices?

There are five principal ways of transporting materials and products through the supply chain: (1) rail, (2) road, (3) pipeline, (4) water, or (5) air. Each of these methods has trade-offs.

Rail

Especially good for shipping bulky cargoes—coal, wheat, automobiles—over long distances on a relatively energy-efficient basis. Railroads handle the greatest volume of domestic goods in the United States, about 40%. They have gone "piggyback," carrying truck trailers on special railcars, and "fishyback," carrying containers from oceangoing container ships.

Road

The most flexible and convenient form of transport because trucks can go almost anywhere and can deliver goods door to door. The major drawback is that trucks are relatively expensive compared to the next three modes of transportation.

Pipeline

A very important kind of conveyance, moving about 20% of the total volume of U.S. domestic goods. Although the products they carry—liquids, such as oil, petroleum, or water, and natural gas—move a slow 3 to 4 miles an hour, they move steadily 24/7, mostly unaffected by weather or labor problems, making pipelines the cheapest mode of transportation.

Water

Ships and barges are the cheapest, but slowest, method of transportation. Besides using planes, there is no other way to transport materials and finished goods to and from overseas points than by ship. However, within the inland waterways and the Great Lakes of the United States, ships and barges also are used to transport heavy, bulky cargoes, such as sand, grain, and scrap metal, as well as standardized containers.

Air

A fast method, but also the most expensive. This makes it appropriate for perishable products, such as seafood and flowers, and items that may be needed quickly, such as medicines. UPS and FedEx have their own fleets of planes for carrying overnight-delivery parcels and cargo, but many passenger airlines also

carry freight cargo in addition to passengers' luggage in their holds. On a ton-mile basis, however, air freight is the most expensive form of transportation.

Intermodal shipping, which combines the use of several different modes of transportation, has become widespread, so that many railroads, for instance, have merged with trucking, air, and shipping companies to provide complete source-to-destination delivery.

This goes hand-in-hand with **containerization**, in which products are packed into 20- or 40-foot-long (by about 8-foot square) containers at the point of origin and retrieved from the containers at the point of destination. In between, the container may, for example, travel by truck to a ship, cross the ocean, be unloaded directly onto a dock, be placed on a railroad flatcar, transported by train across the country, and be unloaded onto a truck for delivery to the final destination.

Comparing Modes of Transportation

Mode	Cost	Speed	On-Time Dependability	Flexibility in Handling	Frequency of Shipments	Availability in Different Locations
Rail	Average	Average	Average	High	Average	Extensive
Road	High	Fast	High	Average	High	Very extensive
Pipeline	Low	Slow	High	Very low	High	Very limited
Water	Very low	Very slow	Average	Very high	Very low	Limited
Air	Very high	Very fast	High	Low	Average	Average

Decisions

I can see this is going to be a big deal as I grow. Right now, Mary, my office manager, has a part-time assistant who handles orders. And we ship them by ground because the bulk of our orders are local.

But that's going to change, right?

If we start selling our products outside our local territory, we need to think about how we will do this. Because time is not a pressing issue, our large bulk orders to the Northwest region distribution center would be most economically moved by _____.

    A. air

    B. water

    C. road

    D. rail

I think you are right, but if customers want our products fast, then we must look at _____ and _____ shipping to meet those needs, even if the cost is much higher, we need to have this option for them.

    A. road; rail

    B. rail; air

    C. water; air

    D. air; road

Correct Answers: D; D

What Are the Differences Between Storage Warehouses and Distribution Centers?

I know I may be putting the cart before the horse, but as my business grows, what are my options for handling inventory?

You really are thinking ahead, Johann. Let's talk about possible choices.

Warehousing is the element of physical distribution that is concerned with storage of goods. Warehouses may be owned by the manufacturer (private warehouses) or be independently owned, storing goods for many companies (public warehouses). The physical handling of goods to and from and within warehouses is called **materials handling**.

Storage warehouses provide storage of products for long periods of time.

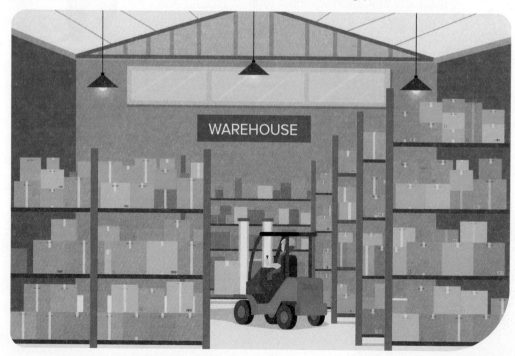

WAREHOUSE

Distribution centers provide storage of products for short periods of time for collection and distribution elsewhere. Companies like Amazon have created a tremendous amount of opportunity for sellers big and small to be able to scale their business through online visibility and managing the shipping and distribution of products.

Decisions

Choose the term that best describes Johann's strategy.

1. Johann wants to get his products on Amazon. He will be sending them large pallets of boxed chocolates that customers can purchase from the site.

   A. Warehouse

   B. Distribution center

2. Johann wants to build up his inventory, so he rents space to hold inventory beyond what his shop can handle.

   A. Warehouse

   B. Distribution center

Correct Answers:1. B; 2. A

Pricing and Distribution: Test

1. The goal of the _____ pricing objective is for the firm to make a profit.

   A. total revenue

   B. gross sales

   C. target return on investment

   D. value target

   E. profit strategy

2. If Portia wants to know how many t-shirts she needs to sell to cover her costs of producing and selling them, she should

   A. create a balance sheet.

   B. run the accounting equation.

   C. analyze her accounts receivable.

   D. create an income statement.

   E. run a break-even analysis.

3. J&R decided to price its lawn equipment to cover the cost of the equipment plus a 20 percent profit. What pricing strategy is J&R using?

   A. demand-based

   B. profit-based

   C. target

   D. cost-based

   E. competitive-based

4. Grace wants to sell her custom jewelry at $25 per piece and needs to make a 15 percent profit. Demand-based pricing determines that her costs must not exceed

   A. $21.74.

   B. $3.75.

   C. $21.25.

   D. $28.75.

   E. $24.00.

5. Columbia sells a men's t-shirt for $19.99 This demonstrates which pricing strategy?

   A. penetration pricing

   B. price skimming

   C. discounting

   D. bundling

   E. psychological pricing

6. _____ is a system for transporting goods or services from producers to customers.

   A. A distribution channel

   B. Materials handling

   C. A delivery system

   D. Conveyance

   E. Supply chain management

7. Which way that intermediaries add value to products occurs when products are made available where it is most convenient?

   A. form utility

   B. location utility

   C. information utility

   D. ownership utility

   E. service utility

8. A(n) _____ wholesaler is owned and operated by the company that makes the products, allowing it to maintain control over the selling and distribution of its products.

   A. full-service merchant

   B. manufacturer-owned

   C. owner-operated

   D. limited-function merchant

   E. drop shipper

9. Morris, Inc. provides products and the means to display them in retail stores and shares the profits with the retailers. Morris, Inc. is a(n)

   A. full-service merchant.

   B. rack jobber.

   C. cash-and-carry wholesaler.

   D. manufacturer-owned wholesaler.

   E. drop shipper.

10. Costco is an example of a(n) _____ because it is a limited-function wholesaler that sells mainly to small retailers who come to the wholesaler and pay cash for a product.

   A. full-service merchant

   B. rack jobber

   C. cash-and-carry wholesaler

   D. manufacturer-owned wholesaler

   E. drop shipper

11. Which retailer is a product-line retailer, as opposed to a bargain retailer?

   A. discount stores

   B. warehouse clubs

   C. outlet stores

   D. secondhand stores

   E. convenience stores

12. Which kind of nonstore retailer sells things like candy and soda through coin-operated machines?

   A. direct selling

   B. direct marketing

   C. vending machine

   D. video marketing

   E. online retailing

13. Sari's Bakery orders 60 pounds of flour and 40 pounds of sugar each week in addition to other ingredients. It also requires packaging for the finished baked goods. Bringing these materials to the bakery involves

   A. inbound logistics.

   B. materials handling.

   C. product management.

   D. physical distribution.

   E. outbound logistics.

14. Ariel's Florist ships fresh flowers all over the United States. What would be the best method of transportation for Ariel's to use?

   A. rail

   B. road

   C. pipeline

   D. water

   E. air

15. Crabapple Candies has fixed costs of $1,000, variable costs of $1 per candy, and price per unit of $4 per candy. How many candies does it need to sell to ensure it breaks even?

   A. 333

   B. 1,000

   C. 250

   D. 200

   E. 334

16. Jamillah runs an automobile repair shop. She thoroughly researches her competitors to analyze their price for oil changes and other simple repairs. She then charges 1 percent less than they do, even when it means lower profits. Jamillah is most likely using a _____ pricing strategy.

   A. break-even based

   B. target costing

   C. competitive-based

   D. psychological costing

   E. demand-based

17. Meti produces homemade candles and is preparing to open a small store to sell her creations. She has analyzed the market and her competitors and decides that she wants to sell her candles for $10.00. Based on her research, she knows she needs to make at least a 10 percent profit on each candle. Using demand-based pricing, we know that Meti's cost to produce the candles must not be greater than

A. $9.09.

B. $11.00.

C. $9.00.

D. $10.09.

E. $10.91.

18. Jitendra works in a large box store. He is always there to greet customers with a smile and ensure that they find exactly what they are looking for. He also believes it is important to answer any questions that the customer might have about the products they are considering. Jitendra is most likely adding _____ utility.

A. retail

B. time

C. form

D. service

E. location

19. Fairway Enterprise owns a wide variety of products but does not maintain the physical goods itself. Instead, Fairway takes online orders from customers and then has the manufacturer ship the product directly to the customer. Fairway is best described as a

A. drop shipper.

B. broker.

C. cash-and-carry wholesaler.

D. direct channel.

E. rack jobber.

20. Lau Technologies manufactures electronics cabling and hubs for consumers. Occasionally, one of its products will be defective and will be shipped back to Lau Technology facilities for analysis and recycling. This example can be best described as a form of _____ logistics.

A. outbound

B. consumer

C. inbound

D. reverse

E. intermediary

What To Expect

By the end of the chapter, you will be able to:

- Construct a promotion mix that meets the three goals of promotion for an integrated marketing communications (IMC) campaign.

- Determine if a push or pull strategy is most appropriate for an integrated marketing communications (IMC) campaign.

Chapter Topics

- **4-1** The Basics of the Promotion Mix
- **4-2** Advertising and Public Relations
- **4-3** Sales Promotion and Personal Selling
- **4-4** Promoting Your Business through Digital Marketing
- **4-5** Developing and Executing a Social Media Campaign

Copyright © McGraw Hill Rawpixel.com/Shutterstock

The Basics of the Promotion Mix

They Love It, But Will They Buy It?

Motivating Consumers to Make Buying Decisions

Nikki is a fitness and fashion aficionado. By day she trains and helps others as a fitness professional. By night she hustles to build her growing line of athleisure wear called Artemis Activwear.

Nikki's customers are busy active women who do it all. They include everyone from hard-working college students, working professionals, and moms. These empowered ladies find time to also invest in their health. They also want to look and feel good while making it happen. Nikki's activewear, which doubles as streetwear, allows them to run errands or tend to the list of to-do's before zipping to her next cardio or barre class in soft yoga pants or sleek leggings, an edgy hoodie framing her face.

Nikki's clients adore her clever and versatile outfits. With a few snips and stitches, Nikki can tweak an ordinary sweatshirt into a stylish top, with an asymmetrical cropped hem, zippered pockets, and lattice cut-outs. She often mixes unexpected items, pairing a boxy fleece jacket with satin ballet shorts. The results are trendy and practical, and people constantly ask her how they can dress as comfortably and stylishly as she does.

Nikki gets permission from the gym manager to set up a rack to display her athleisure clothing line. As clients and members enter, they rush to the rack, holding up items and expressing admiration. Nikki claps her hands.

Hey, everyone. Ready for an awesome workout? Let's get started.

Wait; you brought some of your outfits. What's up?

Just thought I'd show you all what I've been up to lately. Feel free to check it out after class. *(Starts the music; leads the cardio class in the opening warm up. The students turn their backs on the display and follow Nikki.)*

Like most people, Nikki doesn't want to be pushy. She thinks that if her outfits are appealing, they should sell themselves.

After class, she reminds folks to take a look, but many have already rushed off. Only a few stay to look through the rack.

Oh, look, here's a price tag. Not bad. Hey, Nikki, you didn't tell us these were for sale. I love, love, love this top. But it's not my size. Oh well, gotta go; see you at the next class!

Alone in the empty room, Nikki realizes that no one has filled out any orders on the clipboard she had set out. She expected students who praised her Artemis Activwear line to simply write down their customer orders—including quantity, size, and price—on their own.

But now she sees that sales don't happen automatically. Displaying Artemis to potential customers was a start, but she has a lot more work to do to motivate people to actually buy her clothing.

She calls her client Jasmine, a marketing pro who works for a marketing consultancy, to determine how she can better promote her line.

In this lesson, you'll learn how an understanding of the concepts of the *promotion mix*, *integrated marketing communication*, and the *goals of promotion* can help Nikki get the sales she needs.

What Are the Tools, Goals, and Strategies Every Marketer Needs to Build a Promotion Mix

Nikki recognizes she needs some assistance in promoting her activewear brand, Artemis. Nikki enlists the help of Jasmine, a successful marketing consultant and one of her personal raining clients.

Thanks for meeting with me, Jasmine. It looks like the trainer has become the student!

I'm happy to help you with your business. Your classes have always been an inspiration for me. I'm glad to be able to give something back to you.

Well, I think my biggest takeaway from what I've been trying to do so far is that I need to come up with a plan to promote the line.

I would agree with you. We need to have a plan. Word of mouth like you did at the studio the other day is one form of promotion, but there is so much more that we can do. Let's take a look at the the different areas of promotion, and then we can see what is a good fit for your business.

The Promotion Mix

Promotion is used to inform or persuade consumers of the relative merits of a product, service, brand, or issue. It motivates them to buy products. The **promotion mix** is the combination of tools that a company uses to promote a product. These tools should create a unified strategy of integrated marketing communication. These tools are: **advertising**, **public relations**, **personal selling**, and **sales promotion**.

Advertising · Public relations

Promotion mix

Personal selling · Sales promotion

These tools are used to develop a relationship between buyer and seller.

Promotional Tool	Relationship Between Seller and Buyer(s)
Advertising	Paid, nonpersonal communication
Public relations	Unpaid, nonpersonal communication
Personal selling	In-person, face-to-face communication
Sales promotion	Short-term incentives to stimulate consumer buying and dealer interest (e.g., coupons, rebates, trade shows)

Integrated Marketing Communication: A Comprehensive, Unified Promotional Strategy

Integrated marketing communication (IMC) combines all four promotional tools to execute a comprehensive, unified promotional strategy.

Examples that would be incorporated into an IMC strategy may include everything from radio commercials to store-window displays, from print ads to YouTube videos, from direct-mail pieces to T-shirt images—all are designed to present a consistent message from all sources.

For example, a quick-service restaurant launching a new line of healthy menu options may create a campaign that includes:

- a variety of television and digital ads to make customers aware of the new menu options and why the company is adding them to the menu
- billboards, print ads, social media posts, and prominent placing their website all designed to showcase the new items
- sales promotions or discounts that will entice customers to try the new options

Goals of Promotion

Promotion is focused on three main goals, which include:

1. Informing
2. Persuading
3. Reminding

Informing: Telling Prospective Consumers About the Product

Inform people about a product because they will not buy something they know nothing about. Consumers need to be told what the product is, how to use it, where to buy it, and perhaps how much it costs.

Persuading: Inducing Consumers to Buy the Product

Persuade consumers to buy the product, to differentiate the product from competitive products, to say what the unique features are. **Infomercials** are an example.

Reminding: Keeping Consumers Aware of the Product

The last priority is to remind consumers about the existence and benefits of the product.

The Push and Pull of Promotional Strategies

Marketers often use two different promotional strategy approaches to capture the attention of consumers:

1. Push
2. Pull

The **push promotional strategy** is aimed at wholesalers and retailers to encourage them to market the product to consumers. The push strategy focuses on bringing the message and product to a specific customer. As the name implies, marketers are intentionally "pushing" the product to an audience. Marketers may use social and digital marketing efforts to promote the message and other tactics such as samples.

For example, suppose a company is launching a new energy drink. In that case, they may create an Instagram Story or post to broadcast the product launch and pay to have it targeted to their key customer segment's feed.

Additionally, they may send a marketing email to consumers who are already engaged with the brand, letting them know about the launch and where they can acquire the product. Lastly, a company may also and provide free samples to retailers to give to customers.

Often after a marketer "pushes" the outbound marketing message, the following key element is to have a **pull promotional strategy** aimed directly at consumers to get them to demand the product from retailers.

For example, after the launch of the energy drink product, a marketer may utilize the following strategies. They may have influencers write blogs or reviews about the product, post Instagram stories from customers who are fans of the product, and offer a discount code for leaving a review. These approaches can all increase engagement, loyalty, and demand for the product.

Often, push and pull strategies work together because a push strategy helps create demand. A pull strategy helps sales by satisfying the need for the product.

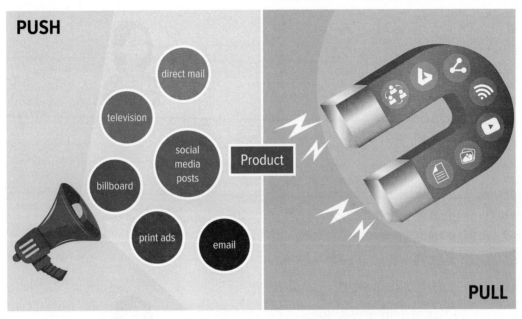

Decisions

Now that Nikki has some background on promotional activities, understands the goals of each, and is aware of a couple of strategies, fill in her dialogue with Jasmine.

You know Jasmine, I realized that I have developed a pretty big number of followers on Instagram. I am thinking about sharing Artemis with those individuals through Instagram, what do you think?

Nikki, this is awesome. You are engaging in _____, the first goal of promotion.

    A. informing

    B. persuading

    C. pull

    D. sales promotion

Yes, I am also thinking about filming a video from the gym talking with a few customers who like the line, and then get some testimonials and post it on YouTube.

Great, Nikki! This knocks out goal two: _____.

A. informing
B. persuading
C. pull
D. sales promotion

Then, I was thinking of providing some short-term incentives via some coupons to stimulate buying.

This is a great idea for _____.

A. informing
B. sales promotion
C. persuading
D. pull

What do you think about asking my followers to ask their gym or local boutique to carry the product? This may help grow retail for the brand.

This would be a great way to start a(n) _____ strategy.

A. informing
B. persuading
C. pull
D. sales promotion

Correct Answers: A; B; B; C

Advertising and Public Relations

By the end of this lesson, you will be able to:

- Determine which type of advertisement is being utilized within a campaign.
- Evaluate the appropriateness of a media plan for an advertisement campaign.
- Evaluate the appropriateness of a media strategy.
- Determine which advertising strategy(ies) is/are being utilized within a campaign.

Generating a Buzz: Advertising and Public Relations

"How Are Advertising and Public Relations Important in Marketing?"

Fitness and fashion expert Nikki is combining her passion and expertise to grow her "athleisure" brand, Artemis Activwear. Helping her on the journey is Jasmine, a client and marketing professional.

Nikki is working hard to build her brand, and Jasmine is guiding her to invest in advertising and public relations.

Nikki, I love your stuff. Seriously, I actually wore it into work the other day and got so many compliments. In fact, one of our partners asked if you'd be interested in sending it to some of our bloggers and celebrity clients.

Wow, no way! Yeah of course, we can definitely do that. I also want to start to use some of the money I am generating to spread the word.

I think you are ready for the next step in growing this brand, which includes advertising and public relations.

Hey, I am ready, coach me up, and let's make it happen!

Nikki wants to grow her brand to reach more customers. Two of the best ways to do this are through advertising and public relations.

In this lesson, you'll learn how an understanding of the types of advertising, advertising strategies, and advertising media, as well as publicity and public relations and how they differ, will support Nikki in her goal to increase product sales.

Types of Advertising: Brand, Institutional, and Public Service

The first of four kinds of promotion, **advertising**, or paid nonpersonal communication by an identified sponsor, uses various media to inform an audience about a product. Three types of advertising are brand advertising, institutional advertising, and public service advertising. **Public relations** is the second kind of promotion.

There are six kinds of advertising strategies: information, reminder, persuasive, competitive, direct action, and fear appeal. Advertising media include newspapers and magazines, television and radio, direct mail, outdoor media, the Internet, and other media, the selection of which involves such considerations as reach, frequency, continuity, and cost.

Whereas advertising is paid coverage of a product, publicity is unpaid coverage. Handling publicity is part of public relations, which is concerned with creating and maintaining a favorable image.

The first area to examine are types of advertising.

Types of Advertising: Brand, Institutional, and Public Service

In general, advertising can be categorized as one of three types: brand advertising, institutional advertising, and public service advertising.

Brand Advertising: Promoting Specific Brands to Ultimate Consumers

Brand advertising, also called *product advertising*, consists of presentations that promote specific brands to ultimate consumers.

This category of advertising is utilized when marketers want to convey a message directly to the consumer. The goal is often to make consumers aware of the brand through direct messaging. By doing this, the company strives to establish credibility and start the process of building a relationship with the customer through deeper engagement.

Example: An auto company may have a new vehicle coming to market and creates an advertisement letting customers know about the new product.

Institutional Advertising: Promoting an Organization's Image

Institutional advertising consists of presentations that promote a favorable image for an organization.

The idea around this advertising category is to be intentional about creating goodwill with consumers. This can also be used to build a desirable image for a company or institution rather than to sell specific products.

Example: "Investing in education. Our idea of a winning formula" is a print ad from computer chip maker Intel promoting the fact that Intel Schools of Distinction rewards programs are proven to raise classroom performance.

One kind of institutional advertising, **advocacy advertising**, is concerned with supporting a particular opinion about an issue.

Public Service Advertising: Promoting Social Causes

Public service advertising consists of presentations, usually sponsored by nonprofit organizations, that are concerned with the welfare of the community in general. Such ads are often presented by the media free of charge.

This category of advertising is sometimes referred to as a PSA. It is primarily used to increase public awareness. It could be used to influence public perception or behavior about a specific topic.

Example: "When you help the American Red Cross, you help America."

Another example: "There's a special joy in getting our hands dirty when it helps keep our land beautiful" is the headline on an ad seeking volunteers to spruce up public lands on National Public Lands Day.

Decisions

Nikki wants to get the most out of her advertising budget. Which strategy best describes the approach Nikki is looking at taking for Artemis Activwear?

1. Nikki believes the best type of advertising is to focus on Artemis, the goddess. She was powerful, being the goddess of the hunt and activity, while also being the goddess of childbirth and mothers.

   On which type of advertising does Nikki want to focus?

   A. Brand advertising

   B. Direct action

   C. Reminder

2. Understanding today's focus on health and well-being, Nikki is considering a strategy that focuses on fitness. She believes that the Ad Council would support and help promote this endeavor.

   What strategy is she considering?

   A. Reminder advertising

   B. Brand advertising

   C. Public service advertising

   D. Institutional advertising

3. For her current customers, Nikki has a mailing list. As new items are released, she is sending her customers email updates.

   What strategy is she considering?

   A. Brand advertising

   B. Direct action

   C. Reminder

   D. Persuasive

4. Nikki is going to use Snapchat with offers for quick sales. What strategy is this?

   A. Reminder

   B. Brand advertising

   C. Persuasive

   D. Direct action

Correct Answers: 1. A; 2. C; 3. C; 4. D

Media Strategies: How to Spread the Word

Advertising permeates the American—and global—business system. It is done by retailers trying to influence consumers, by manufacturers trying to get wholesalers and retailers to carry their products, and by manufacturers trying to get other manufacturers to buy their products. It can be local, as with grocery-store ads; national, as with new-car ads; or cooperative, as when national companies share ad costs with local merchants and wholesalers.

Advantages and Disadvantages of Different Media Types

Advertising media are the ways of communicating a seller's message to buyers.

Media Type	Advantages	Disadvantages
Newspapers	Good market coverage; inexpensive; local market coverage; geographic selectivity	Short life span; limited color options, cluttered pages; declining readership
Magazines	Can target specific audiences; good reproduction and color; long life of ads	Expensive; long lead time; limited demonstration possibilities
Television	Wide reach; good impact; uses sight, sound, and motion; great creative opportunities	High costs of production and air time; short exposure time; short message life; can be skipped with digital recorders
Radio	Low cost; immediacy; good for local markets; can target specific audiences	No visuals; short message life; listeners can't keep ad
Direct Mail	Can target specific audiences; delivers lots of information; ads can be saved	High cost; delivery delays; consumers may reject as junk mail
Outdoor	Highly visible; repeat exposure; low cost; focus on local market	Limited message; general audience
Internet	Inexpensive; interactive; always available; global reach	Consumers may reject as spam

Creating a Media Plan and Strategy

Media planning is the process of choosing the exact kinds of media to be used for an advertising campaign. This means selecting the target audience and determining the best media by which to reach them.

Media Buying: What Are the Considerations?

Often advertisers will come up with a media plan that involves a mix of media, such as having ads on social media platforms, websites, radio, podcasts, and on television. How do you decide which media to buy?

There are four considerations: *reach*, *frequency*, *continuity*, and *cost*.

- **Reach** is the number of people within a given population that an ad will reach at least once.
- **Frequency** is the average number of times each member of the audience is exposed to an ad.
- **Continuity** is timing of the ads, how often they appear or how heavily they are concentrated within a time period.
- **Cost per thousand (CPM)** is the cost a particular medium charges to reach 1,000 people with an ad.

There is also another factor to consider: Digital media, such as social media platforms, have upended the old categories, shifting ad dollars away from traditional media, including TV, newspapers, and magazines.

Publicity and Public Relations

Whereas advertising is paid media coverage of a firm's products, **publicity** is defined as unpaid coverage by the mass media about a firm or its products.

To get the information out to the world, companies issue a **press release**, also known as a *news release* or *publicity release*, which is a brief statement written in the form of a news story or a video program that is released to the mass media to try to get favorable publicity for a firm or its products.

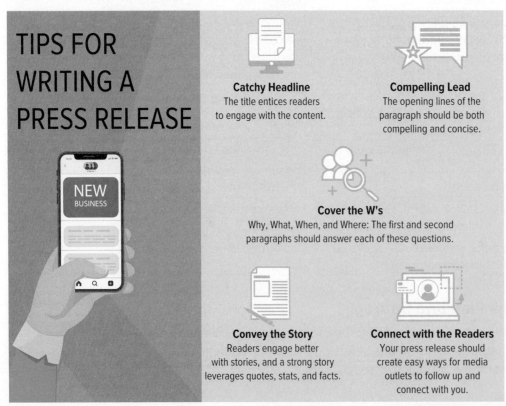

TIPS FOR WRITING A PRESS RELEASE

Catchy Headline
The title entices readers to engage with the content.

Compelling Lead
The opening lines of the paragraph should be both compelling and concise.

Cover the W's
Why, What, When, and Where: The first and second paragraphs should answer each of these questions.

Convey the Story
Readers engage better with stories, and a strong story leverages quotes, stats, and facts.

Connect with the Readers
Your press release should create easy ways for media outlets to follow up and connect with you.

The downside of publicity:

- Treatment (positive or negative) outside of company's control
- No control over whether, when, or where the media will use a news release
- Not likely to be repeated

The process of managing public perception is called **public relations (PR)**, which is concerned with creating and maintaining a favorable image of the firm, its products, and its actions with the mass media, consumers, and the public at large. Public relations often involves:

- Staying connected with public attitudes
- Starting programs in the public's interest
- Informing the public (through the media) about these programs

Decisions

Nikki and Jasmine are planning an advertising and public relations strategy for her business. Fill in the term that best fits the conversation.

Jasmine, I am looking at the cost of some of these media outlets and I don't think I have the budget for it.

Don't worry, Nikki. We can find some great options. Let's first focus on what we want to get out of your media spend.

Well, I want to make a large volume of people know about us, so _____ is really important to me.

    A. press release

    B. cost

    C. reach

    D. Internet

    E. public relations

    F. television

I am still growing and don't have a huge budget, so _____ is important so that we get the most bang for our buck.

- **A.** Internet
- **B.** reach
- **C.** cost
- **D.** public relations
- **E.** press release
- **F.** television

All right, based on that _____ is not realistic because it is high cost, takes production, and has a short message life.

- **A.** television
- **B.** press release
- **C.** public relations
- **D.** reach
- **E.** cost
- **F.** Internet

I agree. I think the _____ is best. We can use Instagram, YouTube, SnapChat, and Facebook Live at a much lower cost, and get to a large audience.

- **A.** cost
- **B.** Internet
- **C.** press release
- **D.** television
- **E.** reach
- **F.** public relations

I agree I think you should also focus on a(n) _____ to hit your key bloggers and authors in the fitness industry. I also have a contact at several women's fashion and health magazines. We can try to get them to write a story about you and the brand.

A. reach

B. Internet

C. public relations

D. television

E. cost

F. press release

Awesome, I did get a direct message from a celebrity. She has an awesome reputation and will be doing some day time talk shows to promote her new movie. She asked if it would be okay to talk about my clothes.

Oh my, that is amazing for _____. You should see if you can send the show some inventory to give to their guests.

A. cost

B. public relations

C. press release

D. Internet

E. reach

F. television

Correct Answers: C; A; B; F; B

Advertising Strategies: From Informational to Fear Appeal

Ads can also be classified by the strategies they take—that is, the approaches or appeals used to try to influence the audience. There are six distinct approaches: **informational**, **reminder**, **persuasive**, **competitive**, **direct-action**, and **fear-appeal**.

Informational Advertising: Providing Straightforward Knowledge

Informational advertising provides consumers with straightforward knowledge about the features of the product offered, such as basic components and price.

For example, an ad from Rosetta Stone, "the world's leading language-learning software," describes such features as its "dynamic immersion" technique, its voice-recognition technology to help with more accurate pronunciation, and its adaptive software that tracks progress and customizes every lesson. It also describes the price and money-back guarantee.

Reminder Advertising: Keeping the Product Visible

Reminder advertising tries to remind consumers of the existence of a product.

For example, Target is a large retail store, and the company may notice a customer's online shopping habits based on what products you purchased from its store and site. Target can then send you an e-mail when those products go on sale.

Persuasive Advertising: Stimulating Desire for the Product

Most advertising is persuasive advertising, which tries to develop a desire among consumers for the product; that is, the ad is not merely informational.

For example, businesses may try to tout the price or value of the product, the product's appealing image, or the product's association with celebrities in its advertisements.

Competitive Advertising: Comparing and Contrasting Products

Competitive advertising, which is also called *comparative advertising*, promotes a product by comparing it more favorably to rival products.

For example, insurance companies GEICO and Progressive each continually run TV ads comparing their prices with those of their supposedly more expensive competitors.

Direct-Action Advertising: Stimulating the Immediate Purchase of a Product

Direct-action advertising attempts to stimulate an immediate, or relatively immediate, purchase of a product through such devices as one-day sales, one-time promotions, or announcements of a special event.

You see direct-action advertising principally offered in e-mail marketing, newspapers, and on local TV channels.

For example, Macy's and JCPenney's have Fourth of July weekend three-day sales, or some commercials and infomercials state customers "who call in the next 10 minutes" will receive a special price or discount.

Fear-Appeal Advertising: Using Worry About Loss or Harm to Sell a Product

Fear-appeal advertising attempts to stimulate the purchase of a product by motivating consumers through fear of loss or harm.

For example, this approach can be used to sell everything from life and auto insurance to fire extinguishers and disaster kits.

Decisions

Now I'm really getting excited to create my ads!

I know what you mean! Because you are creating two ads, what are you thinking of doing for each one?

For the first one, I'm thinking of using the _____ strategy by developing the ad to show exactly what Artemis is all about. Many people don't know our name or even what type of products we offer.

    A. persuasive

    B. reminder

    C. informational

    D. direct-action

Great idea! People really need to be able to identify your brand. It needs to be at the top of their mind when they think about exercise clothing. What about your second ad?

Well, this might be a stretch, but I had another idea of running an ad on social media that shows several news stories about people overheating when exercising due to their clothing choices. People need to know that clothing can be more than just visually appealing, it can be healthy too; and this _____ advertising strategy might be dramatic enough to stick in their minds.

    A. competitive

    B. persuasive

    C. reminder

    D. fear-appeal

I most certainly think you are right that those type of ads will stick in peoples' minds, but are you sure you want to have your potential customers' first impressions of your products be that shocking?

Now that you mention it, maybe I need to wait on that advertising strategy until later. People might associate my clothes with the old cliché ad, "I've fallen, and I can't get up!"

Haha! Yes, that's exactly what came to my mind too! I think your second ad should really make people want to experience your clothing to see if it really works as well as you say it does. How about a(n) _____ advertising strategy in which you offer a sale price coupon code to customers who purchase the clothes online during the upcoming holiday weekend?

    A. informational

    B. competitive

    C. persusasive

    D. direct-action

I love it! I can't wait to get these ads made and watch them appear online! It's fun to think that the ads I create might very well lead to more money!

Indeed!

Correct Answers: C; D; D

Sales Promotion and Personal Selling

By the end of this lesson, you will be able to:

- Evaluate whether personal selling is appropriate in a business case.
- Develop a promotional strategy using B2B, B2C, or other relevant promotion techniques.

Establishing a Relationship Between Nikki and Her Customers

"A Climate of Trust and Intimacy."

Nikki has been growing her athleisure brand, Artemis Activwear. She is amassing a number of customers via hits to her website. She is also gaining interest from buyers who want to put her products in their retail stores.

The challenge is she is not closing the sales. She decides to meet with Dawn, a client she trained who happens to be a VP of product sales for a national department store to figure out how to increase sales.

Nikki, I've got to say you are my hero!

Why is that, Dawn?

Well not only are you building an awesome brand—I know my buyers are really interested, and I am hearing our stores' customers are inquiring about the product—but you also helped me, and many like me, get into the best shape of my life!!!

Thanks, Dawn! I am hoping you can share with me some of your sales wisdom.

You know, Nikki, it really comes down to a few key things: relationships, processes, and execution. I am happy to share with you a few tricks of the trade.

That would be great! Then hopefully I can get a deal with your company!

When Nikki thinks of personal selling, she thinks of pushy salespeople of the sort you might have seen on TV. She doesn't want to be that person. But she realizes that personal selling can benefit both seller and customers by linking them in a relationship of mutual trust. She also sees how systems and processes can make an impact in growing her business.

In this lesson, we'll learn that personal selling, a type of promotion, is face-to-face communication to influence customers. Another type of promotion, sales promotion, is short-term marketing to stimulate dealer effectiveness and consumer buying. Other types of sales promotion include guerrilla marketing and word-of-mouth marketing.

Personal Selling: Establishing a Relationship Between Seller and Buyer

Personal selling is face-to-face communication and promotion to influence customers to buy goods and services. Personal selling involves establishing a personal link between seller and buyer, a kind of "professional intimacy" and climate of trust, particularly in relationship marketing.

It is more expensive than other forms of selling such as telemarketing, direct marketing, and Internet sales because it requires so much training, time, and often money for travel and lodging and, perhaps, entertainment of prospective customers. Thus, this kind of selling isn't appropriate for all situations.

Personal selling works best when the product is:

1. Somewhat expensive or risky to use

2. Somewhat complex and might require the buyer to change behavior

3. Aimed at a few geographically concentrated customers

For example, personal selling often occurs in a business-to-consumer (B2C) situation in which the buyer is purchasing big-ticket items, like a car, diamond ring, or home. Personal selling is also a good approach when products require some explanation, training, or special handling, such as cars, musical instruments, bridal gowns, and life insurance.

Personal selling is also utilized in a Business-to-Business (B2B) context where purchases are typically large and the goal is to cultivate an ongoing long-term relationship with the clients. Some examples could include, farm equipment, office furniture, and medical supplies.

Basic Tasks of Personal Selling

Creative selling is the selling process in which salespeople determine customer needs, then explain their product's benefits to try to persuade buyers to buy the product.

Order processing consists of receiving customer orders and seeing that they are handled correctly and that the product is delivered.

Sales support consists not of selling products but of facilitating the sale by providing supportive services.

Steps in the Personal Selling Process

The personal selling process—by which we mean the creative selling process—consists of a carefully planned sequence of seven activities: (1) prospecting, (2) qualifying, (3) approaching the customer, (4) presenting the product, (5) handling objections, (6) making the sale, and (7) following up. These steps are shown here.

Prospecting

Prospecting: Identifying Potential Customers

Prospecting is the process of identifying potential customers, who are called *prospects*.

First, salespeople must have product knowledge—know the features and prices of their products, as well as those of competitors' products—so they will know what kind of customers they are looking for.

Second, successful salespeople (except those in retail stores) spend a lot of time on preparation—finding their customers and learning as much as they can about them before making their first visit.

Finding customers may be done through referral, endless-chain, or cold-call techniques.

Types of Prospecting

- The **referral sales prospecting technique**: asking satisfied customers to provide names of potential customers and to contact them on behalf of the salesperson.

Example: If you run a home-cleaning service, you might ask some of your most satisfied customers if they would mind endorsing you to their friends.

- The **endless-chain sales prospecting technique**: asking each sales prospect to provide the salesperson with some names of other prospects who might be interested in the product.

Example: A house painter might ask customers for the names of nearby neighbors who might want their homes painted. This technique can be useful even if the first prospect hasn't even bought or used your product.

- The **cold-call sales prospecting technique**: calling on prospects with whom you have had no previous contact and to whom you do not have not any kind of introduction.

Example: If you're selling consulting services, you could simply start searching for companies who may be interested in your services and call their offices to see if you can get someone interested. This is the hardest prospecting technique, of course. It is also apt to be the least successful.

Qualifying

Qualifying: Determining Whether Prospects Have Authority to Buy and Ability to Pay

The second step in personal selling, then, is qualifying—determining if the prospect has the authority to buy and the ability to pay.

Example: If you're selling complex computer systems, you would probably want to ascertain that your prospective client is someone with authority in the information systems department (not, say, the human resources department).

Approaching

Approaching Customers: First Impressions Count

The approach consists of the initial contact with the customer, whether by letter, phone, or personal visit. In most kinds of business-to-consumer (B2C) sales, such as retail-store sales (where most sales take place), not much prospecting and qualifying is required.

With other kinds of sales, such as business-to-business (B2B) sales, or with certain kinds of B2C retail selling such as automobiles or musical instruments, you may need to qualify people as a step in the personal selling process.

Presenting

Presenting the Product: The Canned Versus the Need-Satisfaction Approach

In the sales presentation, you cover your product's features, use, advantages over competing products, and finally the price. You may be aided by free product samples, PowerPoint slides, letters of praise (testimonials) from satisfied users, and a demonstration of how the product is used.

You might approach the presentation in one of two ways—canned or need satisfaction.

- **Canned presentation** uses a fixed, memorized selling approach to present the product. This may be an aid for beginning salespeople, but it's an inflexible approach.

- **Need-satisfaction presentation** consists of determining customer needs and then tailoring your presentation to address those needs. This is by far the more preferred approach these days.

Handling Objections

Handling Objections: Turning Questions into Opportunities

In most sales situations, the prospective customer will raise a few objections, or at the very least have some questions, and you should prepare to have answers for these. Sometimes these can be addressed by introducing others in your company, such as your manager or technical experts.

Frequently, the customer will object to the price, want a discount, or inquire about making monthly payments. Salespeople are better off if they can finish their explanation of the product's features, which may impress the buyer enough that price is less of an objection. In any case, all objections and questions should be viewed as opportunities for strengthening your relationship.

Making the Sale

Making the Sale: The Trial Close and the Actual Close

After you have worked to present all features, answer all questions, and overcome all objections, there comes the time to *make the sale*. Sometimes customers find it difficult to make the final decision, and you have to help them along. Two ways to do this are by making a *trial close* and then the *actual close*.

- **Trial close** is a question or statement that tests the prospect's willingness to buy.
- **Actual close** happens when the salesperson concludes the presentation by asking the prospect to purchase the product.

Following Up

Following Up: Staying in Touch with the Customer

Once the sale is made, you need to do *follow up*—thank your customers for the order, stay in touch with them to make sure they're happy with the product, help solve any problems, and demonstrate that you care about maintaining a long-term relationship. Even retail-store salespeople can benefit from doing follow-up phone calls, especially if they are selling expensive items. Following up is important because it can generate further sales later on or lead to referrals with new customers.

Decisions

Nikki wants to get a meeting with Dawn's national buyer. A deal would mean national retail distribution. Based on the scenario, identify what step she is in in the sales process.

1. Nikki looks over profiles on LinkedIn and determines that Alice is the person who reports to Dawn and ultimately makes the call on what gets purchased.

   A. Presenting the product

   B. Handling objections

   C. Qualifying

   D. Actual close

   E. Approaching the customer

   F. Trial close

2. Nikki asks Dawn to connect her and Alice to set up a meeting.

   A. Handling objections

   B. Actual close

   C. Presenting the product

   D. Trial close

   E. Qualifying

   F. Approaching the customer

3. Nikki has a meeting scheduled at Alice's office where she will share her story, info about the product, and have a demo. The goal is to tailor the pitch to emphasize why Artemis Activwear is a great fit for the shelves of the company.

   A. Qualifying

   B. Handling objections

   C. Approaching the customer

   D. Actual close

   E. Presenting the product

   F. Trial close

4. Alice has some concerns about keeping up with inventory and being able to be at scale. Nikki gracefully lays out her plan and shares her strategy. Alice is pleased with Nikki's preparation.

   A. Approaching the customer

   B. Trial close

   C. Presenting the product

   D. Qualifying

   E. Actual close

   F. Handling objections

5. Nikki presents Alice with several questions to see how interested she is in doing a deal.

   A. Actual close

   B. Qualifying

   C. Trial close

   D. Presenting the product

   E. Approaching the customer

   F. Handling objections

6. Nikki makes an offer based on Alice's feedback, and Alice provides her with a purchase order.

   A. Handling objections

   B. Actual close

   C. Trial close

   D. Approaching the customer

   E. Qualifying

   F. Presenting the product

Correct Answers: C; F; E; F; C; B

Sale Promotions: Techniques Available to Entice Buying

Marketers consistently create new ways to engage and build loyalty with their customers. One common approach is utilizing a point system in which shoppers are rewarded with discounts, cash back, freebies, or a variety of other perks for spending their money and showing their loyalty with a specific company. This is one of many forms of **sales promotion**, which is defined as short-term marketing incentives to stimulate (1) dealer interest and (2) consumer buying. The first is B2B, and the second is B2C.

Business-to-Business (B2B) Sales Promotion

Trade promotion is intended to stimulate dealer interest. The devices used include trade shows, conventions, catalogs, and special printed materials for salespeople. A **trade show** is a gathering of manufacturers in the same industry who display their products to their distributors and dealers.

Examples include makers of furniture, boats, electronics, medical devices, security gear, outdoor equipment, and many other products.

Business-to-Consumer (B2C) Sales Promotion

Business-to-consumer (B2C) sales promotion is extremely varied. Some of the more common devices used are bonuses (such as two products for the price of one), catalogs, cents-off promotions, contests, games, and lotteries. Some businesses even offer free gas to fuel sales. More specialized promotion techniques are shown below.

Specialized Promotion Techniques

Technique	Description
Coupons	Pieces of paper entitling holders to a discount price on a product
Demonstrations	Salespeople show how products work
Event Marketing	Staged, public promotion featuring the product, consumers are active participants in event
Point-of-Purchase Displays	Product displays or ads placed in retail stores where buying decision is made
Premiums	Gifts or prizes provided free to consumers who buy a specific product or accumulate points
Rebates	Potential refunds by the manufacturer to the consumer
Samples	Small product samples given away for free
Sweepstakes	Contests in which prizes are awarded on the basis of chance

Guerrilla Marketing: Innovative, Unusual, and Low-Cost Methods

There are several additional tactics a business can leverage. One such tactic is **guerrilla marketing**, which consists of innovative, low-cost marketing schemes that try to get customers' attention in unusual ways.

For example, individuals entering a sporting event, concert, or attending a conference may encounter a variety of marketers and vendors offering free samples, or promotional items called **swag** to get a customer's attention.

Another example, would be when a company engages in a stunt or creates digital content that goes viral and reaches and engages a wide audience.

Word-of-Mouth Marketing: People Telling Others about Products

Three additional tactics marketers may use include word-of-mouth marketing, buzz marketing, and viral marketing.

- **Word-of-mouth marketing**: A promotional technique in which people tell others about products they've purchased or firms they've used. For example, if you post or share in your social media an experience you've had with a company or shared an unsolicited recommendation with someone then you are engaging in word-of-mouth marketing.

- **Buzz marketing**: Using high-profile entertainment, social media, or news to get people to talk about their product. For example, before a blockbuster movie arrives on the silver screen, studios try to promote the film through trailers and having actors and actresses engage in interviews about the film.

- **Viral marketing**: Companies produce content and, through various channels, the information spreads by being shared and reposted. For example, the ALS Ice Bucket Challenge videos, shares, and reshares was able to raise approximately of $115 million to assist in research, technologies, and care for people with ALS.

Decisions

Nikki is generating some ideas to help grow her business. Choose the term that best applies.

1. Nikki is investigating taking all of her clothing products to a large fashion trade show in Las Vegas next month.

   A. Buzz marketing

   B. B2B

   C. B2C

   D. Viral marketing

2. Nikki is also in the process of visiting all of the local cross-fit gyms in the area to drop off fliers and give out some complimentary T-shirts while recording the give-away live on social media.

   A. Guerrilla marketing and viral marketing

   B. Buzz marketing and going viral

   C. Guerilla marketing and buzz marketing

   D. Viral marketing and word-of-mouth

Correct Answers: 1. B; 2. C

Promoting Your Business through Digital Marketing

By the end of this lesson, you will be able to:

- Describe the three types of media used in online marketing.
- Explain how digital marketing can help a company create customer value, enhance customer relationships, and develop customer experiences.
- Describe how the characteristics of digital buyers and digital sellers influence digital marketing.
- Outline the tactics marketers use to influence online purchasing behavior.

Reaching New Customers Online

Establishing an Online Presence to Grow your Business

Nikki has been growing her athleisure brand, Artemis Activwear. She has established retail relationships, and her product is in several shops. She is also generating sales via her website.

Nikki knows for her company to grow, she needs to expand the awareness of her products and brand. She reaches out to her college friend, Kiera, a social media and digital marketing expert, for some guidance.

Kiera, thanks for taking the time to meet with me. I know that the best next step for Artemis Activwear is to get intentional about our digital strategy. I was hoping you could point me in the right direction.

Nikki, I think you have created a beautiful and functional product. You are spot-on; creating a digital marketing strategy is a great way to expand your brand and ideally grow your sales!

I know you've helped several brands with their digital marketing what do you think is the best way for us to get started.

Thanks, Nikki! I think too often; companies believe digital marketing is just about posting on social media. An effective digital marketing strategy involves understanding the different forms of media and how they create value. You'll also need to know how digital sellers influence digital buyers. Lastly, you'll need to understand the various tools available so you can execute and improve the effectiveness of your digital marketing strategy.

In this lesson, you'll learn the ways in which digital marketing can be utilized to help promote a business to more potential customers online.

The "Big Three" Forms of Online Marketing Media

In today's marketing environment online marketing is key to promoting your business online using a variety of **digital marketing** channels. These channels primarily include search, social, video, email, and display.

Today's consumers are constantly utilizing their devices to engage in everything from researching products and comparing services to digitally bartering and recommending items to friends and strangers alike via social media platforms.

Additionally, audio and video streaming services along with review sites and social media have moved consumer's eyeballs from traditional print media to computers, tablets, and smartphone screens.

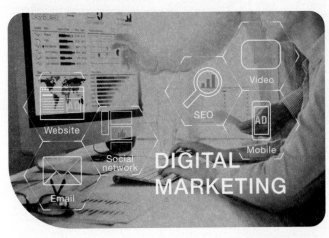

For a marketer, this endless and ever-evolving world of online marketing can seem daunting. However, there are three primary areas digital marketers focus their promotion efforts online to three types of media:

Paid

Paid media comprises of all the online marketing channels that the business pays for such as Google AdWords, Facebook paid ads, and display marketing.

Owned

Owned media is made up of the marketing channels that a company develops such as a website, a customer database that receives your marketing emails, and perhaps a list of readers to an active blog.

Earned

Earned media occurs when a business or company receives recognition or acknowledgment organically. Some examples of earned media would include social media accounts, a mention by a blogger or podcast, or an online article that is written about you, your product, or your company.

In digital marketing, these three media types will have some overlap and interact with one another. For example, you may feature favorable reviews from a site like Amazon (earned media) to your own website (owned media), while paying for a Facebook ad (paid media) to drive customers to your website.

By engaging with customers online through these three areas, a digital marketer has the ability to attract and engage with a variety of customers and potential customers.

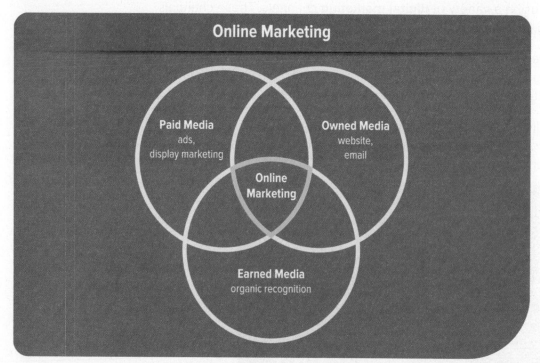

Decisions

Nikki and Kiera meet up to discuss online media marketing.

Nikki Tell me a little bit about how you are currently engaging with customers online?

Well...My biggest presence is through our website and our Instagram account. On our website, we have a blog and offer customers a chance to subscribe to our mailing list which provides them with discounts and early access to our latest designs.

We also have an Instagram account, but we only post on there, we've never purchased and marketing

Lastly, we have been featured in a couple of fitness articles online and I was a guest on a podcast to discuss entrepreneurship.

You are off to a great start, let's look at how you are already leveraging online the three types of online marketing

1. Nikki's website and customer email list are a form of _____ online marketing media

   A. Paid

   B. Earned

   C. Owned

2. While Nikki has an Instagram account she has not purchased any advertising. Keira may suggest that Nikki buy some ads and target them to her target market. Kiera's suggestion would be an example of _____ online marketing media.

   A. Owned

   B. Paid

   C. Earned

3. Nikki's guest appearance on the podcast and the write-ups from fitness sites are examples of _____ online marketing media

   A. Owned

   B. Earned

   C. Paid

Correct Answers: 1. C; 2. B; 3. B

Creating Value Using Different Forms of Digital Marketing

Every company is different; therefore, each company must spend time developing a digital strategy that makes sense for their business.

A company that launches a website, throws up a social media page, and randomly posts content doesn't generate the same results as an organization that invests in creating a focused digital marketing strategy.

The Business and the Audience

When developing a comprehensive digital marketing strategy that adds value for a customer, a marketer needs to understand the business and the audience.

The Business

A marketer has to take a number of variables under consideration when formulating a digital marketing strategy. From a business standpoint, a successful digital marketing professional wants to focus on understanding the following items:

- **The company's mission statement:** helps deliver an authentic presence and it can drive the other components of your strategy.

- **The company's top two to three objectives:** these objectives should be geared toward actions that keep the business going and support the overall identity of the brand.

- **The product or service's value proposition:** What makes this unique, different, or why should a consumer pick your product over another.

- **The company's pitch:** how can the company quickly articulate the business, product, or service to customers. Ideally, this could be articulated in one or two sentences.

The Audience

A successful marketer consistently thinks about the audience. By engaging with the following reflective questions, a marketer can hone an optimal strategy to connect with their intended audience.

Who is the optimal audience?

More than likely, your target market and key market segments are the same consumers you are trying to reach when you deploy traditional marketing efforts.

The challenge in an online setting is that a company should be detailed and specific in identifying the audience because in digital marketing consumers tend to be bombarded with marketing and tend to pursue channels that they strongly market to them.

How do I learn about my audience?

In trying to understand the audience, the focus should be to identify the goals of the particular audience.

Oftentimes, a product or service is a means to an end, so it is important to identify what are the outcomes or end results the customer is really seeking.

Where do I find the audience?

In traditional marketing, we would examine where the customer lives and shop, but in the digital world this is where we need to understand the type of sites they visit, what words they search for, the social media platforms they spend the most time on, what streaming services and apps they engage in most frequently, and so on.

Once a marketer understands the business and the intended audiences, then a specific online marketing strategy can be developed for each audience. A marketer will develop a specific marketing goal, then identify the optimal digital channels that can be used to reach the audience, and finally determine how they will want to use **paid**, **earned**, and **owned media** tactics to execute their online marketing strategy.

Decisions

1. Nikki and Kiera meet up and discuss the mission of Nikki's company, what makes her product unique, and her primary objectives. As it applies to formulating a digital marketing, they are analyzing the _____.

   A. business

   B. audience

   C. product

2. Kiera and Nikki discussing the type of customer who benefits the most from the product, what type of sites they may visit, what social media platforms they are most likely to be on, and how Nikki can learn more about what's most important to them. When developing a digital marketing strategy this type of analysis applies to the _____

   A. business

   B. website

   C. audience

Correct Answers: 1. A; 2. C

Forms of Digital Marketing

When executing a digital marketing strategy, a marketer will typically utilize one or more of the following forms of digital marketing:

Inbound Marketing

Inbound marketing is a form of digital marketing that utilizes such tools as blogging, webinars, or follow-up emails. The intent is to entice consumers to find a firm's products or services without "forcing" an interaction. Inbound marketing attempts to provide value and information to the customer at every stage of the buying journey.

Search Engine Optimization

Search engine optimization (SEO) is the process of driving traffic to a company's website from "free" or "organic" search results using search engines. Try typing in a word or phrase in a search engine and take note of which websites appear first. These companies have utilized SEO to ensure that their sites appear at the top of the list of search results. Thus, SEO is the process of increasing high-quantity and high-quality traffic to a website.

Email Marketing

Email marketing is a cost-effective form of digital marketing that focuses on retaining, nurturing, or attracting customers. The primary advantage of email marketing is that firms know that most consumers have email addresses, which creates a targeted way to reach potential customers.

Social Media Marketing

Social media marketing is one of the most popular forms of digital marketing and is the focus of this lesson. It encompasses marketing activities that utilize online social networks and applications as a method to communicate mass and personalized messages about brands and products.

Decisions

After Kiera and Nikki discuss the different forms of digital marketing, Nikki spends time deciding on which forms will help her optimally engage her audience.

1. Nikki decides to invest more time and money developing her blog. She also starts hosting some fitness and fashion webinars. This is an example of _____.

   A. inbound marketing

   B. search engine optimization

   C. email marketing

   D. social media marketing

2. Nikki decides it is time for her to better leverage the list of email she has and decides to send out a 10% digital coupon to her current customers, and if they refer a friend, they will receive an additional 10% discount. Nikki's goal is to increase her customer list while also retaining and nurturing her current customers. This is an example of _____.

    A. social media marketing

    B. search engine optimization

    C. email marketing

    D. inbound marketing

3. Kiera spends time running Google Ad words and shares with Nikki that she should embed and pay for search words such as "fierce," "fashion," and "fitness apparel". Based on the cost and the return this seems like a promising digital marketing strategy. Kiera is helping Nikki with _____.

    A. inbound marketing

    B. email marketing

    C. social media marketing

    D. search engine optimization

Correct Answers: 1. A; 2. C; 3. D

Creating Value through Website Design and User Experience

For most company's their website serves as a critical element of a digital marketing strategy. A company website serves as the destination for which company's build relationships with customers and in many cases where actual sales take place.

Therefore, it is important that the site creates an exceptional experience that will enhance the customer relationship through content and user experience while delivering a well-designed and highly functional experience that strategically and intentionally makes the customer journey easy and painless.

Ease in Navigation

Users must be able to find what they are looking for in as few clicks as possible.

Quick Load Time

Users expect a seamless browsing experience, so sites must load images and content quickly.

Responsive Design

Because users are engaging with sites with a variety of devices that come in all shapes and forms, it is important to use responsive web design.

By utilizing a responsive design the website will adapt automatically based on the user's screen size, and even the orientation a user is viewing it in.

Customer Service Tools

Companies can offer superior customer service through customer service tools, such as online chat features and the ability to try clothes on a virtual model.

Engaging, Clear Copy

Users want information in as few words as possible. At the same time, the chosen words need to resonate with the user.

Great website copy eliminates long pieces of copy, is not too wordy, and focuses on clear and concise copy.

Also, leverage creative and captivating headings and bullet points to capture your user early and ensure the most important copy is read right away.

The questions to answer when writing great copy:

- Who will read this copy?
- Why are they reading this?
- What emotion do I want them to feel when they read this copy?
- What is the goal of this copy?
- What features and benefits are most important for them to know about our product, service, and company?

Decisions

Nikki recognizes it is time for her to update her website so that her customers have an optimized user experience. While auditing her website she discovered a few tweaks that she could right away.

1. After looking at her analytics Nikki realized that customers were dropping off and it was clear there are simply too many clicks for customers to make. Nikki works with her web designer to ensure customers have as few as clicks as possible when they visit her website. Nikki is addressing _____.

   A. ease in navigation

   B. responsive design

   C. customer service tools

2. A recent report came out showing that most Nikki's Target Market prefers to browse and shop from their mobile device. Nikki realizes that her website needs to be adjusted to show up better on different mobile devices. Nikki is addressing which of the following elements of her website?

   A. ease in navigation

   B. customer service tools

   C. responsive design

The Influence of Digital Buyers and Sellers

It is important to recognize buyers come to our sites for different reasons and at different stages of the consumer purchasing process. Meanwhile, companies have different approaches and strategies for selling their products and services online. Therefore, it is important to understand the different types of digital buyers.

Digital Buying Behaviors

Just as in any market environment, different segments of buyers exist in the digital environment. Understanding which buyers tend to turn to digital retail can help marketers create compelling and effective digital marketing campaigns.

Online purchasing behavior falls into five distinct types: product-focused shopping, browsing, researching, bargain hunting, and one-time shopping.

Product-Focused Shopping

Consumers who engage in **product-focused shopping** either replace an existing product or purchase a product chosen for them—a textbook, for example. This type of shopping might include regularly repurchased household staples such as paper towels and shampoo.

Browsing

Consumers that engage in **browsing** behavior are not really looking to make a purchase. Browsers tend to fill shopping carts that they later abandon, although some may reactivate those carts at a later date. They will typically browse their favorite sites just for fun, kill time, or look for new ideas.

Researching

Consumers that engage in **researching** behavior are looking to buy a product for the first time. Unlike browsing, which has no expected outcome, researching is more deliberate and will result in a purchase online or offline.

Bargain Hunting

Consumers that engage in **bargain hunting** behavior usually explore coupon sites such as wish.com or auction sites such as eBay. Bargain hunting is often combined with browsing and may or may not lead to a purchase.

One-Time Shopping

Consumers that engage in **one-time shopping** behavior may combine product-focused shopping, browsing, researching, and bargain hunting all at the same time. Such consumers are shopping for a gift or using a gift card and will typically not return to the shop once the purchase is made.

Digital Sellers

Just as there are various buyer segments in the digital retail space, there are a variety of digital sellers. The most common are digital malls, digital marketplaces, auction sites, and buying club sites.

Digital Mall

A **digital mall** is where a variety of sellers stock their goods. When you click on an item, you can be directed to several purchase options, each with different prices and terms, such as free or expedited delivery. In a digital mall, consumers are often unaware of which firm is actually selling them the goods.

For example, if you click "purchase" on a small local business website, you may be redirected to Amazon.com, which handles the actual sale. Amazon.com (similar to Alibaba.com) is an example of a digital mall.

Digital Marketplaces

A **digital marketplace** is made up of small, independent sellers—Etsy.com is a good, global example of this type of online retailer. Consumers search Etsy for a variety of handcrafted or vintage goods sold by a variety of sellers.

Auction Sites

An **auction site** such as eBay lists goods from individuals or firms that can be purchased through an auction bidding process or directly through a "purchase now" feature.

Buying Club Sites

A **buying club site** such as Boxed.com allows consumers to buy in bulk, similar to warehouse clubs such as Sam's Club or BJ's.

When engaging in online marketing, it is important to keep in mind; several consumer segments could be visiting the same digital sites for reasons other than to make a purchase. Therefore, as part of a digital marketing strategy, it is important to analyze site visitor behavior to identify which segments are most common and develop strategies that will help convert these customers into sales.

For example, through user tracking, marketers can determine how many times the average consumer visits a site before purchasing. Studying visitor behavior can also tell firms what types of activities these consumers engage in (are they researching or browsing?) and which pages they tend to visit.

Decisions

Kiera and Nikki spend some time analyzing her digital buyers and identifying some digital selling strategies that may help grow her business.

1. About 40% of Nikki's customers are avid about their fitness, and know exactly the function and fashion they are looking for in their apparel. Which of the following best describes this consumer?

   A. Browser

   B. Product-focused shopper

   C. One-time shopper

   D. Bargain hunter

2. Another 30% of digital buyers in Nikki's product category know what they are looking for and are ready to purchase, but they aren't necessarily brand loyal, so they tend to look at quality, price, and deals. They also shouldn't be counted on as return shoppers.

   A. Browser

   B. One-time shopper

   C. Product-focused shopper

   D. Bargain hunter

3. Based on the analysis of the 70% of the digital buyers, Kiera suggests that Nikki should expand her digital presence beyond her website.

Because part of Artemis's brand strategy is to be known as an established high-quality product, Nikki wants to avoid sites where people can bid on her product.

She also recognizes her goal is not to be a small independent seller. She wants to get the most exposure and be featured where other big brands in the category are being sold. Based on Nikki's goals and strategies she should pursue a(n) _____.

A. digital mall

B. buying club site

C. auction site

D. digital marketplace

Correct Answers: 1. B; 2. B; 3. A

The Tools Online Marketers Use to Influence Purchasing Behavior

Digital marketers spend a considerable amount of time engaging with data and analytics to understand their customers' purchasing behaviors.

Based on the data and analytics, marketers utilize various tools to influence customers by making them aware, piquing their interest, increasing their desire, and ultimately getting the person to take action and make a purchase. When it comes to online purchasing, marketers influence consumers' behavior in both visible and invisible ways.

Visible Ways Marketers Pay for Influence

Digital marketers have the opportunity to invest resources into several categories to reach customers and influence customer behavior. The most common paid digital marketing tools include: paid search, paid stories, paid display advertising, and sponsorship.

Paid Search

Paid search is online advertising in which a company pays to be a sponsored result of a customer's Web search. Companies usually purchase a given search term—"best business school," for example—and pay each time a customer clicks on their sponsored search ad. Paid search advertising is usually sold using an auction system with the search terms awarded to the highest bidder.

Paid Stories

Paid stories are ads that appear as content designed to look like stories to the viewer. Most advertisers and publishers aspire to deliver paid ads that are cohesive with the page content.

For example, on a page discussing new technology products, a visitor may see a "story" about a brand with the latest smartphone targeted to people who want the latest technology. In this example, the brand would have paid for the "story" to appear on the page.

Paid Display Advertising

Paid display advertising includes everything from banner ads to YouTube video advertising. These ads generate awareness as well as (hopefully) drive traffic to a website. Companies can also purchase advertising on social media sites such as Instagram and Facebook.

Sponsorship

Sponsorship firms may compensate or contract with an individual who has a large number of followers on social media platforms such as YouTube, TikTok, or Instagram to endorse, showcase, or demonstrate the firms' products. These individuals are often referred to as **influencers** as their endorsement of a product can influence their followers to potentially make a purchase.

For example, a celebrity may be paid a fee to discuss the firm's products on their YouTube channel. The influencer may receive additional compensation if a consumer clicks on the display ad via the YouTube page.

The Invisible Tools Marketers Use to Influence Behavior

Digital marketers also leverage several tools that consumers may not notice or be conscious of when trying to influence purchasing behavior; these often include cookies, geotracking, and bots.

Cookies

Cookies are small data files stored on websites that can generate a user profile about a consumer; the profile might include browsing history or login information.

For example, have you ever searched Google for a product and then noticed that the next time you opened your browser or a social networking site like Facebook, the ads that appear on the page were for the product you searched for? How did those ads get there? The answer is with cookies.

Consumers are not typically aware of these cookies and the data or "crumbs" they leave behind and provide at each site they visit. *Cookie synching* allows cookies to embed in a person's computer and follow that person as they move through different websites.

Through cookies, marketers know what consumers are searching for and where they are searching.

Geotracking

Geotracking allows marketers to know a consumer's geographic location. Based on this information can be used to push paid ads or sales promotions such as discounts.

For example, Target may push notifications about a sale on a product if geotracking shows that competing stores are also in the consumer's local area.

Bots

Bots are a software application that runs automated tasks over the Internet. On Facebook and other websites, bots look for keywords such as "tickets," "NFL," and "hotel," and then automatically send consumers information related to those keywords. These marketing messages might show up as an ad on Facebook or a sponsored ad on Instagram. Or, consumers might receive an email touting a special deal on tickets for their team's upcoming game.

Through these online tools, marketers can provide consumers with content that will be most impactful and influential when engaging in the consumer decision-making process.

Decisions

Kiera and Nikki examine some opportunities to invest in digital influence for Artemis Activwear.

1. Nikki decides she wants to invest some of her digital marketing budget to the term "best women's sportswear." This is an example of which of the following paid influence?

   A. paid display advertising

   B. paid search

   C. paid stories

2. Kiera shares with Nikki that a very popular fitness Instagram user was wearing Nikki's clothes in her latest post. Kiera recommended reaching out to the individual to see if she would be willing to partner with Nikki by reviewing and endorsing the product. In turn, Nikki would provide the influencer with a special link to Nikki's website and the influencer would receive a commission for products purchased via the link. This is an example of _____.

   A. sponsorship

   B. paid stories

   C. a paid search

3. Kiera recommends Nikki use _____, which are small data files stored on websites. She believes this will help Nikki because this behind-the-scenes digital marketing will help get a better return on their Facebook advertising because it will target individuals who visit Nikki's site with meaningful Artemis Activwear advertisements.

   A. cookies

   B. geotracking

   C. bots

Developing and Executing a Social Media Campaign

By the end of this lesson, you will be able to:

- Describe the ways social media fosters communication between companies and consumers.
- Discuss the ways market professionals utilize mobile marketing to engage with customers.
- Summarize the steps used in developing a social media campaign.
- Outline the four steps used in executing a social media campaign.

Connecting with Customers Online

Leveraging Social Media to Grow Your Business

Nikki has been growing her athleisure brand, Artemis Activwear. She has been working with her college friend, Kiera, a social media and digital marketing expert, to build a robust digital marketing strategy. Nikki has a done an excellent job of implementing several digital tools and expanding her company's digital presence. However, she recognizes she needs to invest further in her social medial and mobile marketing. She meets up with Kiera to discuss the next steps.

Hi Kiera, your insight has been so valuable as I develop Artemis's digital marketing plan.

No problem, Nikki! I am excited to help you explore social media and mobile marketing.

Yes, I know it takes time, effort, money, and a little luck to deploy a thriving social and mobile strategy.

You are right, but with your work ethic and commitment, focus on learning and growing will make all the difference.

I appreciate it! With that, let's go to work!

You got it! Let's dive in by looking at customer interactions, feedback, and some different mobile applications. By starting here, we will build, execute, and evaluate a solid social media campaign strategy.

In this lesson, you'll learn how to promote a business to more potential customers online through social and mobile applications.

Creating Connections through Social Media

One of the key benefits of utilizing social media campaigns as part of a digital marketing strategy is that companies can offer customizable, targeted, and two-way communication with customers.

For example, a company may post information about a new product that provides consumers the opportunity to provide feedback in various ways, such as likes, re-tweets, shares, comments, and so on. The company can then leverage this information to make decisions around their marketing mix for the product. These unique qualities of social media compared to that of traditional mass media make **social media platforms** effective, flexible, and adaptable tools for marketers.

Consumer Interactions

Social media platforms have utilized technology to deliver a personalized experience to consumers as each person receives different content through their social feeds. Social media platforms include Facebook, Snapchat, Instagram, LinkedIn, TikTok, and Twitter.

This allows companies to be specific in targeting their message to key customer segments. This is different than promoting through traditional media because marketers were typically only able to deliver mass messaging via television, radio, and print.

Social media allows marketers to create interesting and creative messages and content aimed at specific targeted markets. The challenge becomes captivating the audience because consumers are inundated with a large volume of messages. Social media can provide a forum for consumers to broadcast their thoughts to a large audience. Typically, this occurs in three key ways online.

Customer Feedback to the Company

Social media allow consumers to interact with the company without actually having to visit a store or talk to a company representative. Typical ways to interact with a company include via websites, videos, podcasts, and social media pages. Some interactions include purchasing products, while others include requesting service, asking questions, or simply conducting informational research.

Peer to Peer Interaction Online

In addition to company interaction, social media also allow consumers to interact with each other. This type of interaction used to occur through personal discussions called "word of mouth," where consumers described to each other both good and bad experiences with a company or product. Today, social media allow this type of interaction to easily occur through blogs, forums, panels, reviews, and comments.

Content Creation

Social media also allow consumers to interact with people indirectly, which can contribute to content creation. Many consumers enjoy voicing their opinion of products, sports teams, companies, and many other things. Typical ways that consumers create content are through blogs, posts, videos, pictures, podcasts, and status updates within social media sites like Instagram, TikTok, Pinterest, and Facebook.

Consumer Feedback

A key feature of social media is that consumers have the opportunity to easily react, respond, and provide feedback in real time to companies.

Consumer feedback consists of different ways to report their satisfaction or dissatisfaction with a firm's products. Consumer feedback can take many forms:

Consumer Reviews

A common form of consumer feedback is provided through consumer reviews. A **consumer review** is a direct assessment of a product (good, service, or idea) expressed through social media for others to see and consider. Choosing what book to read, a podcast to listen to, where to eat, or what hotel is best is heavily influenced by consumer reviews.

Chatter (comments, likes, retweets, shares, hashtags, etc.)

Often a marketer's message is further broadcast to other consumers. This practice, called **chatter,** occurs when a consumer shares, forwards, or "retweets" a marketing message. Chatter is another form of consumer feedback. For marketers, the level of chatter represents consumer feedback.

Influencers

Consumers who engage with social media at higher levels often become influencers. **Social media influencers** are consumers who have a large following and credibility within a certain market segment.

Influencers are often bloggers, celebrities, industry experts, or even people who produce relevant YouTube videos. Those who have a strong following will often be asked to report on new products as a high-visibility form of feedback.

Many applications exist that allow companies to monitor all consumer feedback in one consolidated location. Products such as Crimson Hexagon, Facebook Insights, HootSuite, Klout, and Mention are cloud-based software applications that specialize in bringing together a company's social media feedback in one place.

Decisions

Nikki and Kiera meet up to go over some of Artemis Activwear's consumer interactions and feedback.

1. On Nikki's website, she allows customers postpurchase to provide a rating along with a review of the product. Overall, she has a 4.8-star rating out of 5. This type of customer interaction would be considered which of the following?

   A. Customer feedback

   B. Peer-to-peer interaction

   C. Content creation

2. Nikki and Kiera have developed a calendar for Nikki to post to her most followed social media accounts. Kiera recommends that Nikki post videos that will help people improve their physical, mental, and emotional fitness and motivational videos of people working out in the apparel. This type of customer interaction would be considered which of the following?

    A. Peer-to-peer interaction

    B. Customer feedback

    C. Content creation

3. Kiera suggests that Nikki start asking individuals to use the #ArtemisFit in her posts, as well as to share, like, and retweet. The reason is because Kiera wants Nikki to leverage her customers overwhelming positive feedback to grow brand awareness. This is an example of _____.

    A. influencers

    B. consumer reviews

    C. chatter

Correct Answers: 1. A; 2. C; 3. C

Deploying Mobile Marketing Tools

Mobile devices are playing an increasingly important role in consumer's daily habits. Therefore, company's develop mobile marketing efforts to communicate with customers on their devices. Specifically, **mobile marketing** is a set of practices that enables organizations to communicate and engage with their audience through any mobile device or network.

Social Media and Mobile Marketing

The technology used to develop mobile devices creates a delivery tool that allows marketers to utilize other digital strategies such as social media as part of their mobile marketing efforts. Mobile marketing uses many of the same platforms as social media marketing.

For example, mobile marketing allows marketers to know your geographic location and push your messages with deals, promotions, and discounts.

Similar to social media, mobile marketing provides customers the opportunity to engage with companies in real time.

For example, you can share a review of a restaurant as you are eating dinner, or you can evaluate a ride-share driver's performance as soon as you get out of the car. In some cases, a company may push or text a survey to you right after receiving a service to get your immediate reactions.

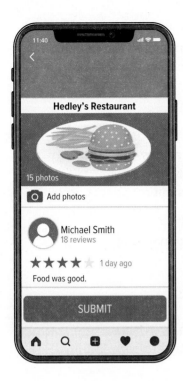
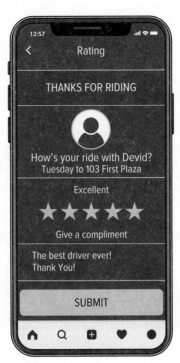

Mobile Applications Used in Social Media

Marketers use different types of mobile applications in their social and mobile marketing campaigns. They choose which applications to use and which delivery platforms are most efficient (e.g., mobile, website, etc.) based on which will most appeal to their target audience. Review the examples in the nearby table to better understand the benefits of using traditional social media platforms versus a company-specific application.

Mobile Application Types

	Examples	Benefit
Traditional Social Media Platforms	Facebook, Twitter, Instagram, TikTok, LinkedIn, and Snapchat	Reach a broad audience of consumers across various mobile apps and Wi-Fi-based website technologies.
Company-Specific Applications	Chipotle, Domino's Pizza, Nordstrom, and Target	Reach a specific audience and gather data for key customer segments. Integrate features such a geotargeting, loyalty, mobile payments, and user experience

One of the keys to success in online marketing is for a marketer to consistently be thinking about opportunities to improve the consumer's online experience.

Digital marketers can utilize applications to help create a positive online experience. While there are numerous applications, most companies can utilize the following applications to communicate, connect, and improve the customer experience.

Geotargeting

Many firms incorporate **geotargeting** within their apps. This technology allows firms to know when customers (who have the related apps) are physically near their stores. Marketers use geotargeting to push marketing messages to consumers within a store's vicinity to entice them to it.

In terms of services, firms like Uber have used mobile technology's locational ability to revolutionize the taxi industry.

Loyalty Programs

Many companies manage **loyalty programs** through mobile applications. Starbucks has its own successful app that tracks users' loyalty points and allows users to pay for their morning coffee through the app.

In fact, 30 percent of all Starbucks' sales last year were made through its mobile app. Starbucks' mobile app also notifies users of deals, promotions, and ways to earn more loyalty.

Mobile Payments

Mobile technology makes it easier for consumers to pay for products. Companies are becoming increasingly willing to accept **mobile payments** via Apple Pay or Android Pay, eliminating the need for consumers to carry cash or credit cards. This also includes applications such as PayPal, Venmo, and Zelle that allow individuals to pay peers and vendors. Also, many companies offer gift cards that are mobile-based, making it easy to track balances and make purchases.

Scheduling

Scheduling apps allow service companies like hair salons, gyms, and restaurants to be convenient for their customers to schedule appointments or manage reoccurring services.

For example, at Great Clips hair salon, customers who check-in online are granted priority status over those who do not.

Information

Mobile apps are also quite useful for simple things, such as information.

Movie theaters, airlines, auto dealers, and almost all businesses can provide great deals of information as consumers desire it, even when they are on the go.

Decisions

Nikki and Kiera look at some different ways Nikki can expand her social and mobile presence.

1. Nikki would like to experiment with a specific ad on YouTube when a person is in proximity to a big box gym. This idea is an example of _____.

   A. geotargeting

   B. mobile payment

   C. loyalty program

2. Kiera suggests that Nikki start allowing individuals to pay on her site with Venmo and PayPal. This strategy would be a good utilization of _____.

   A. geotargeting

   B. loyalty program

   C. mobile payment

Correct Answers: 1. A; 2. C

Developing a Social Media Marketing Campaign Strategy

A **social media marketing campaign** is a coordinated marketing effort to advance marketing goals using one or more social media platforms.

Social media marketing campaigns share many similarities with traditional marketing campaigns. Both require setting goals, choosing strategies, and iterating on the plan as the campaign evolves.

However, in many ways, social media campaigns differ from traditional marketing campaigns. Let's look at the steps involved in developing a social media marketing campaign.

STEP 1	STEP 2	STEP 3	STEP 4
Set a Goal	Select the Correct Combination of Social Media Platforms	Create Marketing Content	Monitor the Campaign

Step 1: Set a Goal

The first step in a social media marketing campaign (as with any marketing campaign) is setting a goal. For example, the goal may be to increase brand awareness, increase the number of app downloads, or communicate product information. To achieve their goals, marketers develop strategies. To choose the right strategy, marketers start by defining what area of the marketing mix (the four Ps) they will be addressing with their social media and/or mobile campaign.

Step 2: Select the Correct Combination of Social Media Platforms

The next step is to select the correct combination of social media platforms. To accomplish this, marketers must determine their target audience. Sometimes the most appropriate platform is the company's own website, whereas other times the best platform is an external social media platform that has a larger reach with the target audience. Often, a combination of social media and websites is used at the same time.

Step 3: Create Marketing Content

Next, marketers create content for the audience. **Content** is the information, images, videos, and any other delivery method of the marketer's social media message.

Step 4: Monitor the Campaign

Finally, the results of the campaign are monitored. In Step 1, goals were set for the campaign; in this step they are used to monitor how successful (or unsuccessful) a social media campaign is. If the plan isn't working, changes will need to be made. Tourism officials in Egypt, for instance, created a social media marketing campaign with the hashtag #thisisegypt. However, the campaign backfired when people used the hashtag to show the negative realities of living in Egypt. Because this social media campaign delivered results counter to what was intended, it clearly needed adjustment.

Decisions

Nikki is excited to meet with Kiera so she can develop a new social media marketing campaign for Artemis Activwear.

> Hi Kiera, for this campaign, I want to increase brand awareness, and with my new line launching this month, I think this is the perfect direction for our social media marketing campaign
>
> 1. Which step in developing a social media marketing campaign is Nikki referring to?
>   A. Monitor the campaign
>   B. Create marketing content
>   C. Select the correct combination of social media platforms
>   D. Set a goal

> That's awesome, Nikki! Based on the analytics we look at and the fact you'll want to showcase the new line, I think that leveraging TikTok, Instagram, and Facebook will be the best applications for the campaign.
>
> 2. Which step in developing a social media marketing campaign is Kiera referring to?
>   A. Monitor the campaign
>   B. Set a goal
>   C. Select the correct combination of social media platforms
>   D. Create marketing content

> Yes, your idea around building our brand with the #ArtemisFit hashtag has given us some great posts, videos, and interactions to use in the campaign. I will also be working with a few influencers, and we'll be developing a schedule with video releases up until the launch.
>
> 3. Which step in developing a social media marketing campaign is Nikki referring to?
>   A. Set a goal
>   B. Monitor the campaign
>   C. Select the correct combination of social media platforms
>   D. Create marketing content

That sounds fantastic, and great job leveraging the audience you have been building! My only other recommendation would be to watch your analytics and some of the customer feedback and make some adjustments to maintain the message and impact you want to have with the campaign.

4. Which step in developing a social media marketing campaign is Kiera referring to?

  A. Select the correct combination of social media platforms

  B. Create marketing content

  C. Monitor the campaign

  D. Set a goal

Correct Answers: 1. D; 2. C; 3. D; 4. C

Executing a Social Media Campaign

Once a social media campaign strategy has been developed, there are typically four steps to execute the campaign.

Step 1: Set a Goal—Integrating Social Media into the Marketing Mix

Social media relates to all four areas of the marketing mix. In different ways, social media can be used as a place (to sell products), product (typically a service), promotion (communication method), or price (communication method).

Product	Place
Social media and mobile technology can be used together as a means to deliver service products.	Social media is integral to the place of the marketing mix by being with consumers wherever they go.
Example	In essence, a marketer's "store" can be anywhere. Using social media, consumers can shop, share, and interact with companies from wherever they are.
You can use Airbnb to book a place to stay, Rover, to find a place for your pet to stay, and Uber to take you to and from your desired destinations or deliver a meal from your local restaurant.	**Example**
	Marketers also capitalize on the strengths of mobile technology and social media to improve traffic at their physical locations.

Promotion	Price
Using social media is a powerful way to educate consumers about current promotions, both online and in-store. These promotions are spread by social media users as well as by the company.	Social media have also changed the way that companies price their products. Stores are adapting to customer's ability to quickly price compare by offering price matching and making price adjustments based on customers' comments online.
Companies try to generate "Chatter" by trying to create a **viral campaign,** in which messages spread quickly by social media users forwarding promotional messages throughout their social networks.	Companies are also expanding the methods in which consumers can pay consumers to pay for products by allowing social and mobile apps for transactions.
Example	**Example**
A powerful example of a viral campaign is the ALS ice bucket challenge in which individuals raised awareness and money for ALS research by pouring ice-cold buckets of water on themselves then challenging others to do the same. The results increased research funding by 187 percent.	For example, an increasing number of stores accept PayPal, Apple Pay, and Google Pay as methods of payment, and some businesses are now allowing Venmo for services.

Step 2: Select the Correct Combination of Social Media Platforms

The choice of which social media platform a marketer chooses for a particular strategy is driven by the goals and target market of the strategic marketing plan. Marketers have a wide variety of choices to implement their plan. They start by asking three initial questions when considering which social media to use for a specific marketing campaign.

Where is our target audience spending their time on social media?

This is the largest determiner of the social media site choice. Placing marketing messages in the right place for the right audience is a fundamental marketing principle.

What type of content do we want to create within the campaign?

If marketers want to create engaging videos, YouTube may be a good choice, whereas Instagram is well suited to still-image campaigns. If the campaign is designed to share professional articles and news about the company intended for a professional audience, then LinkedIn may be an appropriate channel.

What social media channels are our competitors using?

Often, maintaining a marketing presence in proximity to their competitors is an effective way to highlight product differentiation. Marketers learn a lot about what customers want and how that is changing by observing what their competitors are doing with their social media.

Step 3: Create Marketing Content—Set Social Media Objectives

In social media marketing, "Content Is King"; therefore, it is imperative that the images, copy, and video are aligned with the messages you want to deliver to customers.

- Social media content stimulates conversation among consumers. The conversation generated through the campaign can amplify and spread the message.
- Social media platforms customer engagement and serve as a prime location for promoting products, building brand awareness through effective content, and further developing relationships with customers by sparking conversations.
- Additional social media marketing goals revolve around determining the needs of customers and increasing customer service.

While a company sets overall marketing goals, it is important to establish the objectives, or actions that need to be taken to create results and achieve the goals.

For example, a company may establish a goal of increasing sales of a new product they are about to launch. Therefore, they may establish that they need to reach a certain number of potential customers in a target market and they will do that by creating five posts a day for a week leading up to the launch and post them on Instagram, and spend a certain dollar amount on Facebook ads to reach their intended audience.

Step 4: Monitor the Campaign

The goals and objectives of social media marketing campaigns can be measured using many tools that are provided by the social media sites themselves. In addition, other applications have been developed for this purpose, such as Google Analytics or Viralheat, which allow marketers to manage and monitor multiple social media accounts in one central location through the use of a **dashboard.**

The dashboard aggregates (counts up) the number of likes, shares, pins, clicks, views, comments, and other variables in one spot, so social media marketers can quickly assess how well they are implementing their objectives and what progress is being made toward their goals.

Dashboards can also provide other important information, such as the peak times when people are conversing about their brand. Companies like Tableau provide solutions to businesses to visualize data in real-time.

To monitor the success of a campaign, companies can determine bounce rates, click paths, and conversion rates through various online sites:

Bounce Rate

A **bounce rate** is the percentage of visitors who enter a website and then quickly depart, or bounce, rather than continuing to view other pages within the same site.

In this way, bounce rates measure how many people leave a page immediately after clicking on an ad.

Click Path

A **click path** is a sequence of hyperlink clicks that a website visitor follows on a given site, recorded and reviewed in the order the consumer viewed each page after clicking on the hyperlink. Click paths show marketers how customers interact with a page once they click on the ad and can help the company streamline the process from first click to purchase.

Conversion Path

A **conversion rate** is the percentage of users who take a desired action, such as making a purchase. Conversion rates tell marketers how many people who click on an ad actually complete a purchase or subscription or follow the company.

Evaluating the Success of a Social Media Marketing Campaign via Reach, Frequency, and Sentiment

Reach and Frequency

Many companies also try to quantify their likes and shares based on the profit they generate. To make this calculation, companies determine how many consumers they are connecting with (**reach**) and how often they connect with these consumers (**frequency**).

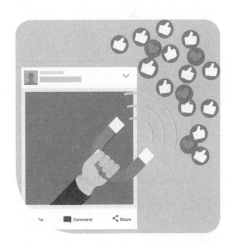

The reach and frequency statistics are then compared to the expense of all social media activities taken to achieve these results. However, firms cannot perfectly match reach and frequency with dollar amounts because determining reach, frequency, and exact expenditures to produce them is often difficult.

Thus, more effective measures of financial return can be gleaned from examining where customers come from. To determine where customers come from, firms can track what social media message immediately preceded a website visit that resulted in a purchase.

For example, if you receive a Twitter message that a popular band will be coming to your area and then click a link to purchase a ticket, it is clear that you "came" from a Twitter message. Because most firms use multiple social media platforms to communicate, determining which one drives the most visits and purchases is useful for planning social media campaigns.

Sentiment Analysis

Marketers also measure how consumers are responding to their content by using sentiment analysis. **Sentiment analysis** looks at whether people are reacting favorably or unfavorably to products or marketing efforts.

Sentiment analysis examines all posts and comments on all streams to come up with sentiment. Different software systems are capable of examining sentiment, with most categorizing sentiments as either positive, negative, or neutral.

The advantage of sentiment analysis is that it can be run continually throughout the campaign, allowing a company to make adjustments to its strategic plan as needed.

Decisions

Nikki's and Kiera connect to discuss the impact of her social media marketing campaign

1. Nikki's goal was to increase brand awareness as she prepared to launch her latest product line. Nikki shares that she saw an increase in traffic to her website, sales were up compared to her last launch, and she had an extremely low _____ percentage, meaning customers stayed on her webpage after clicking on the ad.

   A. bounce rate

   B. conversion path

   C. click path

2. In addition to examining sales and digital analytics, Nikki was also pleased with the campaign's _____, which showed overly favorable reactions to both the content and the new product line.

   A. conversion path

   B. sentiment analysis

   C. frequency

   D. reach

Correct Answers: 1. A; 2. B

Promotion Mix: Tools, Goals, and Strategies: Test

1. The _____ is the combination of tools that a company uses to promote a product.

   A. marketing mix

   B. promotional toolkit

   C. promotion mix

   D. marketing package

   E. promotional package

2. John Doe is running for a congressional seat in Texas, and he feels his best demographic is businesspeople who work along the Beltway in Houston. What would be the best media option for John to reach this target market?

   A. outdoor

   B. television

   C. direct mail

   D. radio

   E. magazines

3. An ad for a drug for acid reflux explicitly explains what ingredients are in the drug, how it works to reduce reflux, and what the possible side effects are. Which advertising approach does this ad use?

   A. informational

   B. reminder

   C. control

   D. competitive

   E. direct-action

4. After Evan closed the sale of window replacements, he asked the customer for names of potential customers in the neighborhood. This is a way to conduct

   A. qualifying.

   B. approaching.

   C. prospecting.

   D. making the sale.

   E. presenting.

5. Kendall works second shift and always watches television for a few hours before he goes to sleep. He came across a lengthy commercial that advertised a multitool that could do the work of over 1,000 other tools. The woman in the commercial provided demonstrations of the tool in many real-life scenarios. When the commercial was over, Kendall went online and purchased one of the multitools for himself. This commercial can be best described as achieving which of the following goals of promotion?

   A. persuading

   B. informing

   C. branding

   D. pushing

   E. reminding

6. Sterling is asked to appear on the Channel 3 morning show to discuss his company's corporate social responsibility initiative. Which part of the promotion mix does this best represent?

   A. public relations

   B. sales promotion

   C. brand advertising

   D. corporate selling

   E. advertising

7. The Federal Emergency Management Agency (FEMA) and the Ad Council worked together to create a television commercial stressing the importance of weather radios and home disaster kits. This commercial can be best described as _____ advertising

   A. public service

   B. competitive

   C. institutional

   D. brand

   E. reminder

8. Ruby Healthcare ran a series of print ads in fitness magazines that included the tagline, "Dedicated to Building the Health of Our Members and Our Communities." These print ads can be best described as _____ advertising.

   A. fear-appeal

   B. institutional

   C. public service

   D. direct-action

   E. brand

9. Dog Time Toyz has recently hired a vice president to create and build up its corporate giving and volunteerism programs. The marketing team creates a short bio regarding the new vice president's background plans for the department before sending it to major news outlets for publication. This bio can be best described as a(n)

   A. press release.

   B. media plan.

   C. institutional release.

   D. publicity stunt.

   E. advocacy campaign.

10. Dion is part of the marketing team at Stafford Shipping. He organizes a meeting with his team to review and select the media sources they will be using for next year's advertising. This activity can be best described as

    A. personal selling.

    B. brand continuity.

    C. media planning.

    D. trade promotion.

    E. public relations.

11. Franco identified a list of five potential customers. After reaching out to the five individuals, he asks each one for the name of five other leads. He then repeats the process with the new group of leads. Which of the following tasks in the personal selling process does this best represent?

    A. approaching

    B. creating

    C. qualifying

    D. requesting

    E. prospecting

12. Valu-Rite places a campground arrangement in the center of its convenience store each summer and surrounds it with chocolate, marshmallows, and graham crackers—everything needed for s'mores. Which of the following promotion techniques does this best represent?

    A. point-of-purchase displays

    B. demonstrations

    C. event marketing

    D. guerrilla marketing

    E. samples

13. The person who is most likely to make a purchase is the consumer who is

    A. browsing.

    B. product-focused shopping.

    C. bargain hunting.

    D. price comparing.

    E. seeking customer support.

14. After typing "women's swimsuits" into her browser, ads for swimsuits began to show up in Tyra's Facebook feed. This is a result of

    A. cookies.

    B. geotracking.

    C. sponsorship.

    D. paid display advertising.

    E. search engine optimization.

15. A YouTube sensation who produces videos about fashion is provided clothing samples from different brands because of her large following of young people. In this scenario, the YouTube personality is best described as a(n)

    A. product guru.

    B. opinion leader.

    C. social networking star.

    D. social media influencer.

    E. online persuader.

16. Which is essential to selecting the correct combination of social media platforms for a social media marketing campaign?

    A. determining the target audience

    B. deciding what content will be used

    C. understanding product pricing

    D. figuring out what content will create the most chatter

    E. developing social media objectives

17. Tanheya works for a real estate firm that specializes in vacation properties. She is in the process of revamping the firm's social media campaign because she has determined that its existing one is not reaching its intended audience. In what step of the social media marketing campaign is Tanheya's firm?

    A. setting goals for the campaign

    B. selecting the correct combination of social media platforms

    C. monitoring the campaign

    D. creating the delivery method for the social media message

    E. integrating social media into the mix

18. To meet the goals of a social media campaign, certain actions, or _____, need to be taken to create results and achieve the goals.

    A. expectations

    B. deliverables

    C. promotions

    D. determinations

    E. objectives

19. In which scenario would click paths be most helpful to a marketer?

    A. The marketer wants to know how many people who click on the ad complete a purchase.

    B. A company wants to streamline the process from first click to purchase.

    C. The marketing department wants to find out how many people leave a page after clicking on an ad.

    D. The social media plan calls for an analysis of the demographics of those who are reacting to the company's marketing efforts.

    E. The firm wants to understand the customer's entire journey from social medial platform to purchase.

20. A pet accessories company learns that people are buying more of its pet beds after featuring a series of videos showing kittens and puppies fast asleep and cuddling together in these beds. This finding is reflective of a(n) _____ analysis.

    A. sentiment

    B. feedback

    C. interest

    D. reaction

    E. favor

5 Introduction to Marketing

What To Expect

By the end of the chapter, you will be able to:

- Recall how marketing adds value to an organization's goods and services.
- Differentiate among products, product lines, and product mixes.

The Basics of Marketing

Marketing: America's Pastime

"How Can We Take Advantage of Marketing Opportunities?"

Mark is about to start a summer marketing internship program with the minor league baseball team, The Great Falls Golden Doodles. "The Doods," as they are affectionately known around town, plan to engage Mark in several marketing tasks as they prepare for the upcoming season.

It is Mark's first day of work, and although he is a little anxious, he is also excited to get started and help the team however he can!

Welcome to the minors, Rookie! I'm Jim, the general manager. We are thrilled you are joining us for this season. This is Ginger; she is our marketing manager.

Yes, Mark, we're excited to work with you. Because our business is primarily based on the season, we HAVE to make sure we capitalize on our key marketing opportunities.

So, before the season is literally in full swing, we think it's important for you to gain an understanding of our "value-add" and the products we sell here at the ball park—or as our fans affectionately call it "The Dog House."

Let's get your stuff put away and get you started.

In this lesson, we'll help Mark add value to the team.

Marketing is a core function of business. It helps businesses understand the needs, wants, and desires of customers. This understanding is then used to develop, price, distribute, and promote value-added products and services in the marketplace. Marketing evolved over four eras: the production, selling, marketing concept, and customer relationship eras. Value is an important part of marketing. Companies market products, product lines, and a product mix.

Understanding Marketing

According to the American Marketing Association, **marketing** is the activity, set of institutions, and processes for creating, communicating, delivering, and exchanging offerings that have value for customers, clients, partners, and society at large. Marketing is practiced by both:

For-Profit Organizations	**Nonprofit Organizations**
We see marketing being practiced everywhere by profit-oriented firms to deliver goods (Tide, Chipotle, and Chevrolet) and services (Hilton, Verizon, and Netflix).	We also see marketing used more and more by nonprofits, whether private-sector organizations such as Harvard University, the Presbyterian Church, the Red Cross, or the American Cancer Society, or public-sector organizations such as the University of Nebraska, the U.S. Postal Service, or the U.S. Marines.

Companies strive to use a **marketing concept**, which focuses on customer satisfaction, service, and profitability.

Customer Satisfaction: "We Need to Give Buyers What They Expect from Us."

Learning what customers want and giving it to them isn't a unique idea, but it's surprising how many firms emphasize promotion or sales instead. **Customer satisfaction** is the concept of offering a product to please buyers by meeting their expectations. The most successful American businesses consistently stay aware of what customers want and provide it to them.

Focus on Serving Customers: "Everyone, from CEO to Stock Clerk, Should Focus on Customer Service."

Firms that successfully employ the marketing concept integrate their approach so that everyone in the organization—from top management to entry-level associates—focuses on the same goal of satisfying the customer. This is why you may see managers at local supermarkets step up and help with bagging groceries during busy periods at the checkout counters. The "not my job" attitude doesn't work here.

Emphasis on Profitability, Not Sales: "We Need to Concentrate on the Products That Are Most Profitable."

Successful firms focus on offering the goods and services that are most profitable, not on offering the entire range of products and not on total sales. For example, a restaurant may have only a handful of menu items that drive revenue and profitability, therefore they may choose to revise their menu to focus on what items keep them in business.

The marketing concept combined with the advent of digital marketing and social media has led to a further refinement known as the relationship management concept. **Customer relationship management (CRM)** emphasizes finding out everything possible about customers and then using that information to satisfy and even exceed their expectations in order to build customer loyalty over the long term.

Delivering Value and Building Relationships with Customers

Companies focus on utilizing marketing to produce and deliver offerings that have value—for customers, clients, partners, and society at large. **Value** is defined as a customer's perception that a certain product offers a better relationship between costs and benefits than competitors' products do.

Note that we used the word *perception*—it is not the actual value of one product compared to another, but rather how the customer perceives that value.

Decisions

Jim and Ginger meet to help Mark get more acquainted with the role marketing plays in the organization. Ginger explains to him the role marketing plays on the team.

Mark, our strategy is to utilize _____ to focus on customer satisfaction, service, and profitability.

- **A.** a marketing concept
- **B.** the value-added process
- **C.** the profitability concept
- **D.** the customer relationship process

Because season ticket holders are very important to our success, we utilize a _____ system to find out as much information as we can about them to ensure we are exceeding their expectations and building long-term loyalty.

A. marketing

B. CRM

C. production-era

D. value

We've heard that we need to enhance our food menu and have better apparel options for our fans. We believe this will help us increase the perceived _____ of the fan experience.

A. marketability

B. value

C. profitability

D. marketing concept

How to Distinguish Products, Product Lines, and Product Mixes

Whatever the type of marketing, the point is to get consumers to buy or use the organization's products. Product-oriented organizations may carry some sort of product(s), have product lines, have a product mix, or some combination of the three. As a company grows and evolves they may go from a single product to offering a number of products so as to better survive wide swings in demand.

Product: A Good or Service That Can Satisfy Buyers' Needs

A **product** is a good (which is tangible) or service (intangible) that can satisfy customer needs.

Examples: A product can be almost anything: goods such as tomato soup, motorcycles, hearing aids, or houses, or services such as auto insurance, plumbing repair, Internet connection, hotel stay, or college education.

Product Line: A Group of Products Designed for a Similar Market

A **product line** is a collection of products designed for a similar market or that are physically similar.

Examples: Campbell Soup Company sells not only tomato soup but also a product line of other condensed soups: mushroom, minestrone, and so on. State Farm offers not only auto but also life, home, and health insurance.

Product Mix: The Combination of All Product Lines

A **product mix** is the combination of all product lines that a firm offers.

Examples: Campbell offers condensed soups, Supper Bakes meal kits, Prego Italian sauces, Swanson broth, and Pepperidge Farm cookies.

Decisions

Mark, you're learning very quickly. And I could really use your help with a few tasks today. In my upcoming meeting with Jim, he wants us to understand what our core products, product lines, and product mixes are going to be for the season. I've written them down, and I'd like for you to go through the list and put each in the appropriate category.

1. Which of the following lists best represents a product line?

   A. Hot dogs, hats, tickets, baseball bats, big foam fingers

   B. Party planning or Uber rides to and from events

   C. Regular Coke, Coke Zero, Diet Coke, Coke Life, and Cherry Coke

   D. Hot dogs, burgers, nachos, and our specialty items

   E. Tickets as single-game, small packages, and season tickets

2. Which of the following best represents a product mix? Choose all that apply.

   A. Hot dogs, brats, and polish sausages

   B. Hot dogs, burgers, nachos, and our specialty items

   C. Tickets as single-game, small packages, and season tickets

Correct Answers: 1. C; 2. B, C

Consumer Buying Behavior

By the end of this lesson, you will be able to:

- Categorize customer behaviors into the steps of the consumer buying process with which they are associated.
- Differentiate among the factors that influence buying behavior.

Consumer Buying Behavior: Ball Game or Movie?

Mark is quickly learning that marketing plays a major role in the success of an organization. He understands that it is imperative for the business to add value to its customers. Therefore, marketers must learn as much as they can about customers and how they make their purchasing decisions.

You've seen some of the ways we generate revenue beyond ticket sales. Now, we'd like to get your help in better understanding our fans.

We want you to utilize the consumer buying behavior process to help us understand how we can better serve our fans. Last year, we noticed a dip in attendance, concessions sales, and revenue from our team gift shop.

Yes, we have a crucial goal to increase the revenue that's generated.

Let us know what you find based on some of this data we collected from fans.

Copyright © McGraw Hill

For this task, Mark will need to help the Doods better understand their customers. In this lesson, we'll help him learn a few things about how consumers make decisions and what influences their buying behavior.

Deciding What to Purchase: The Consumer Buying Process

Consumers don't automatically make a purchasing decision. In fact, they typically work through a process to determine whether they will ultimately make a purchase. By understanding how the consumer thinks and what influences their process, marketers can more effectively communicate and assist consumers in that decision-making process.

How Does the Consumer Buying Process Work?

The **consumer buying process** consists of five steps by which consumers make decisions: problem recognition, information search, evaluation of alternatives, purchase decision, and postpurchase evaluation.

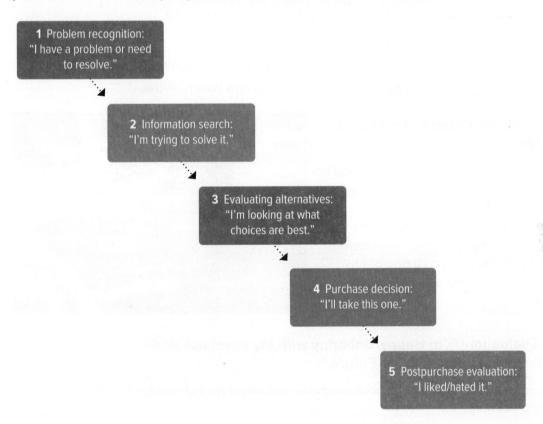

1 Problem recognition: "I have a problem or need to resolve."

2 Information search: "I'm trying to solve it."

3 Evaluating alternatives: "I'm looking at what choices are best."

4 Purchase decision: "I'll take this one."

5 Postpurchase evaluation: "I liked/hated it."

Problem Recognition: "I Realize I Have a Problem to Resolve."

Here you discover you have a problem or a need that needs to be addressed, such as hunger.

Example: You are wrapping up a long day of work, and you know there are limited food options in your fridge at home.

Information Search: "I Need to Find a Solution to My Problem."

Here you do some sort of search for a solution to your problem. For example:

- You could go to the store and buy ingredients, or a premade meal.
- You could stop at a fast-food or quick service restaurant on your way home.
- You could eat at a restaurant.
- You could put in an order to a restaurant through an app such as UberEats and have food brought to you at home.

Evaluating Alternatives: "I'll Weigh the Pros and Cons of the Products Available."

After gaining information on competing products, you consider the benefits and drawbacks of each. For example:

- You are pretty tired and don't necessarily have the energy to stop at the store, buy ingredients, and make something, nor do you really feel like going to a restaurant and sitting down.
- You want something a bit higher quality than fast food.
- You don't mind spending a little extra today to get something you'll enjoy.
- You want to get home and enjoy your meal while you relax and watch Netflix.

Purchase Decision: "I'll Choose This One or Not Choose at All."

Finally, you make your decision. You decide you'll order food from one of your favorite restaurants through UberEats. You don't mind spending a few extra dollars on the service because you know you'll get a good quality meal from one of your favorite restaurants, and you'll be able to time the delivery to arrive shortly after you get home from work.

Postpurchase Evaluation: "I'm Happy/Unhappy with My Purchase and Will/Will Not Buy a Similar One in the Future."

Marketers hope you will be satisfied with your purchase and might be inclined to repeat your choice in the future.

On the other hand, you might suffer unhappiness, *buyer's remorse*, for having bought something that you perceive later to be too expensive, too shoddy, lacking other products' features, not delivering on its promises, and so on.

You may decide while this was a good decision for this particular situation, having food delivered more than once a week would be too expensive and not a good use of your income.

Decisions

Mark, did you have a good weekend?

I guess it was okay. Nothing spectacular, but come to think about it, I'm not really sure what I did to tell you the truth.

Haha, same here. It got me thinking about why so many people like coming to the watch the Doods. They are probably bored with life, and this is a way to make them feel connected to something. This is the concept of _____, and at least they are taking steps to resolve it.

    A. personal satisfaction

    B. evaluation

    C. problem recognition

You know, if you think about it, you're right! What if we try to work that into our advertisements? What if we try to give people a sense of belonging here at the park watching *their* Doods play each week?

Great idea! It's almost like you are getting inside the heads of our customers and figuring out _____.

    A. customer buying behavior

    B. the product life cycle

    C. psychological factors

That's exactly what I'm doing, I think? Yeah, that's what I'm doing! When customers start _____ of what to do with their time, what else do they have to choose from that is even close to what the Doods can offer them on a weekly basis?

    A. recognizing the problem

    B. evaluating alternatives

    C. taking corrective action

Haha, well, Mark, you might be overinflating the impact of a minor league baseball team on a person's life, but you really are making a point about why consumers desire certain products. They want to feel like they are solving a problem they have at the moment, and maybe we should think about making people want to feel part of something cohesive here at the ball park. The Doods might just be what they need!

Correct Answers: C; A; B

Factors That Influence Buying Behavior

Throughout the buying process, consumers are influenced by various factors, from global to personal, from general to specific. It is important for marketers to learn and become aware of these factors as they impact what consumers may or may not purchase.

Culture and Subculture

The Influence of Values and Attitudes

Consumers' ideas about ways of doing things are passed along to us from earlier generations of whatever nationality we are (culture) and the ethnic, religious, age, education, gender, and other groups we belong to (subculture). These influence our buying decisions.

Consumers may want to come to a baseball game because it is considered the national pastime, because they were once part of team, or because the community around them supports the team.

Examples of culture and subculture: nationality, gender and age, ethnicity and race, and religion

Social Class

The Influence of Our Socioeconomic Group

Our choices are also affected by our social class, whether lower, middle, or upper. The cars and houses people buy, for instance, reflect their income levels.

In minor league baseball, we can create an experience at a price for different income levels allowing fans at all socioeconomic levels to enjoy coming to a game. We do that through our ticket pricing, price of apparel, concessions offerings, and a wide variety of products and services.

Examples of social class: income level, profession, educational level, social status, and political status

Reference Groups

The Influence of Groups We Identify With

We are also affected by those special groups we belong to or identify with: family, friends, fellow students, coworkers, music lovers, fraternity/sorority members, and so on.

If you're an athlete, for example, you may favor the kind of footwear worn by other athletes, rather than, say, those in country music bands. This is one of the great parts of being a fan, you feel connected to the team and can feel like you are part of the team by cheering with other fans, wearing the apparel, and joining friends, family, and coworkers at a game.

Examples of reference groups: family, friends, fellow students, college groups, and coworkers

Personal Image

The Look We Wish to Project

A lot of us want to project a certain image, based on the products we buy.

This is why marketers often recruit sports stars and other celebrities to promote products. As an organization, teams want to project an image that plays to a wide array of people so that coming to a game is the thing to do!

Examples of personal image: physical look, profession, lifestyle, and fashion

Situational Matters

The Effects of Timing, Moods, Impulse, and Expectations

All kinds of other things affect our buying decisions: timing, moods, impulse, expectations, advertising, pricing, and beliefs about a product.

Teams try to create situations with special promotions, incentives for coming to the game such as discounts and giveaways from sponsors. It also helps when the team is winning!

Examples of situational matters: timing, coincidence, mood, and impulse

Decisions

Mark, our _____ here at the ballpark has been on higher-level income fans, but I think we need to find a way to attract a larger fan base including people with lower incomes.

- **A.** subculture
- **B.** geographic
- **C.** reference group
- **D.** socioeconomic focus

I noticed that while being at the first couple games. It seems to me that if the Doods are only appealing to one type of person, then we have an opportunity to attract the _____ they identify with, but we will probably miss out on others.

- **A.** classes
- **B.** marketing mix
- **C.** reference groups
- **D.** people

Yes, and that concerns me too, since we could be appealing to a much larger group of people in general. Let's run a few targeted promotions based on social media by offering a deep discount if the tickets are purchased within 10 minutes of the ad running. Maybe this will entice certain groups of people to buy the tickets on a whim, and maybe these _____ will attract them.

- **A.** situation matters
- **B.** subcultures
- **C.** consumer preferences
- **D.** reference groups

I'll get right on it! I'm curious to see if this will work.

Correct Answers: D; C; A

Lesson 5-3

Marketing Strategy

By the end of this lesson, you will be able to:

- Differentiate among the three approaches to market strategy.
- Differentiate among the types of consumer segmentation.
- Differentiate among the types of business segmentation.

Building a Winning Marketing Strategy

As the season approaches, the front office team needs to build a winning marketing strategy that will help them target the right customers. Mark, Ginger, and Jim huddle to discuss how to create a marketing strategy.

Wow, Mark, you are proving yourself to be an MVP in the marketing department!

We've been really impressed with the work you've done so far. You're definitely ready for your next task.

Last season, we were able to capture some great information about our fans. This season, we want to use that information to develop a target market profile and to help us better understand our key market segmentation.

Once we know this information, we'll be able to develop a better marketing strategy for the upcoming season.

And we'll have the right marketing mix to deliver value to our fans!

In this lesson, we'll help Mark execute this task with the team, by learning about how markets are classified, what constitutes a marketing strategy, and the different ways to segment a market.

Developing a Strategy for a Target Market

A marketing strategy is a plan for identifying the target market among market segments (groups), creating the right marketing mix, and dealing with the external environment. Marketers targeting consumers may market to geographic, demographic, psychographic, benefit, or user-rate segments and may also resort to niche marketing and one-to-one marketing. Business-to-business marketers classify business markets into geographic, customer-based, and product-use-based segments. Nonprofit marketers draw on five types of marketing approaches: people, place, event, cause, and organization.

Marketing, which begins with learning who your customers are and what they need and want, may be of three types:

- For-profit marketing to consumers
- For-profit marketing of businesses to other businesses
- Nonprofit marketing

How Do You Begin to Develop a Marketing Strategy?

It starts with—this is important!—having a **marketing strategy**, a plan for (1) identifying the target market among market segments, (2) creating the right marketing mix to reach that target market, and (3) dealing with important forces in the external marketing environment. Buyers or users of a product fall into all kinds of groups, or segments.

Thus, a marketing strategy relies on

- **Market segmentation**, dividing a market into groups whose members have similar characteristics or wants and needs.
- **Target marketing strategy**, consisting of marketing directly to such segments—the target market. Marketers don't have endless resources. Thus, they need to direct their efforts toward people who are most likely to buy their products.

Marketing professionals look to create a well-defined target market strategy by defining the following consumer attributes.

Decimal

Decisions

Mark, since you've been at many of the games, what are your thoughts about the type of customer the Doods appeal to the most?

You know, I have been thinking about this, and I have noticed a large part of our audience seems to be women in their 20s and 30s. It's strange because I thought I would notice more men.

Good eye, Mark! You have nailed our _____.

A. target market

B. marketing strategy

C. mass market

D. attributes

Really? Well, I guess I'm that good!

The Doods seems to be attracting a younger female audience, so we have been thinking about developing a full _____ to target this group.

A. marketing strategy

B. for-profit marketing concept plan

C. checklist

D. market attribute plan

True, and I think we could _____ the market even more with a bit more research to help us really narrow down on this group.

A. mass

B. divide

C. segment

D. attribute

I think you're right! Grab your notebook and binoculars, and let's meet at the game tonight at 7:00 PM!

Correct Answers: A; A; C

Market Segmentation

The **consumer market** consists of all those individuals or households that want goods or services for their personal use. These are all the products you buy every day, from toothpaste to insurance to watching Netflix.

Consumers are divided into five segments by **geographic segmentation**, **demographic segmentation**, **psychographic segmentation**, **benefit segmentation**, and **user-rate segmentation**.

Geographic

Dividing the Market by Location

Geographic segmentation categorizes customers according to geographic location.

Examples: A nationwide retailer of hiking boots would tend to advertise more in Colorado (with its Rocky Mountain trails) than in New York City. Makers of tortillas would promote their products more in Texas (with its large Hispanic population) than in Minnesota.

Demographic

Dividing the Market by Age, Gender, or Income

Demographic segmentation consists of categorizing consumers according to statistical characteristics of a population, such as gender, age, income, education, social class, ethnicity, and so on.

Example: Clothing is segmented first by gender, then by age, then by income level.

Psychographic

Dividing the Market by Psychological Characteristics, Values, and Lifestyles

Psychographic segmentation consists of categorizing people according to lifestyle, values, and psychological characteristics.

Examples: Frugal versus free spending, or rebel versus conservative.

Benefit

Dividing the Market by Benefits That People Seek in a Product

Benefit segmentation consists of categorizing people according to the benefits, or attributes, they seek in a product.

Examples: Style versus economy, or safety versus speed, or high-tech versus low-tech features.

User Rate

Dividing the Market by Frequency of Customer Usage

User-rate segmentation consists of categorizing people according to volume or frequency of usage, as with heavy users versus light users.

Example: 60% of U.S. adults purchased wine in 2013; however, the Wine Market Council reported that just over one-third (35%) of adults aged 21 years and older consumed wine in 2013.

Taking Segmentation Even Further: Niche Marketing and One-to-One Marketing

Segmentation can be taken even further through two other processes: (1) niche marketing and (2) one-to-one marketing.

Niche Marketing: Dividing Marketing Segments into Microsegments

Niche marketing consists of dividing market segments even further to microsegments for which sales may be profitable.

For example, men who grow long beards may be actively seeking and purchasing grooming products for their beards, but those products would not necessarily have appeal outside the niche.

One-to-One Marketing: Reducing Market Segmentation to Individual Customers

One-to-one marketing consists of reducing market segmentation to the smallest part—individual customers. Marketers need to do intensive research to gain a deep understanding of a customer's preferences and keep detailed records on customer interactions.

For example, common uses are online recommendations for books and movies by Amazon.com based on consumers' previous purchasing or viewing histories. High-end applications are sales of expensive real estate, boats, and cars, where a salesperson may collect all kinds of information about a wealthy prospect, then craft a custom sales pitch.

For-Profit Marketing to Businesses: Goods and Services for Business Use

The **business market** or **business-to-business (B2B) market**, also known as the industrial or organizational market, consists of those business individuals and organizations that want business goods and services that will help them produce or supply their own business goods and services.

What Are Three Market Segments That Businesses May Target in Marketing to Other Businesses?

Business markets can be classified into three categories: geographic, customer-based, and product-use-based.

Category	Description	Example
Geographic	As with the consumer version, the business version of geographic segmentation consists of categorizing customers according to their geographic location. Industries are often grouped in certain geographical areas.	The part of the San Francisco Bay area near Palo Alto that has been dubbed Silicon Valley hosts such well-known information-technology companies as Google, Yahoo, Facebook, Apple, Intel, Oracle, and Sun Microsystems.
Customer-based	Resembling demographic segmentation for consumers, in the business market, customer-based segmentation consists of categorizing business customers according to specific characteristics.	Examples might include size, industry type, and product/service-related attributes.
Product-use-based	Product-use-based segmentation categorizes business customers according to how they will use the seller's product.	A manufacturer of GPS (global positioning system) devices might divide its target market into, say, long-haul trucking companies, taxi-cab companies, delivery companies (pizza, flowers), home health care services, security companies, and so on.

Decisions

Mark, were you able to make any sense of the information about our fans that I gave you last week?

That was a lot of information, and I tried my best to break it down into categories just to get a grip on it.

Good! That is actually called _____, and it is the first step in the right direction.

A. market segementation

B. mass marketing

C. niche marketing

D. customer breakdown

See, my schooling is paying off, right?

I would say it has! Did you use _____ segmentation to divide the market up by things such as age, gender, religion, and so on?

A. geographic

B. psychographic

C. demographic

D. personal

I sure did, and then I also dove into where most of the fans are driving in from to go to the games.

Excellent, I was just about to ask you if you _____ segmented your results.

- **A.** categorically
- **B.** psychographically
- **C.** demographically
- **D.** geographically

When I compared the two together, I see we have our largest fan base coming from within 10 miles of the stadium, and they are women, who are about 28 to 40 years old.

Fantastic! I have been trying to figure out who our primary target actually is, and I do believe we are getting close to being able to find a better way to market directly to this group!

Correct Answers: A; C; D

Lesson 5-4
The 4 Ps

By the end of this lesson, you will be able to:

- Recall the reasons a firm may develop a new product.
- Recall the four key strategy considerations (the 4 Ps) of a marketing mix.

The Marketing Mix

"Let's Explore Product, Price, Place, and Promotion."

A marketing plan comes together when marketers deploy the 4 Ps of marketing: product, price, place (distribution), and promotion around their target market. Ginger will need Mark's assistance to begin to construct what is known as a marketing mix that applies all 4 Ps to a target market ensuring the right products are offered at the right price, and distributed and promoted to the customer in the optimal manner.

"Don't look back," said legendary ballplayer Satchel Paige. "Something might be gaining on you." This sentiment also reflects an important reason to produce new products, but it is only the first of four reasons.

And those are?

Don't worry, Mark. We'll discuss those in more detail later on. But understand that this information becomes the basis for determining the marketing mix, which consists of four key strategy considerations—the "4 Ps" of product, pricing, place, and promotion.

In this lesson, we'll help Mark and Ginger create marketing mixes for the Doods.

Companies have to develop new products for four important reasons. Such development requires doing initial research and identifying the target market. This information becomes the basis for determining the marketing mix, which consists of four key strategy considerations—the "4 Ps" of product, price, place, and promotion.

Four Reasons to Develop New Products

A **new product** is defined as a product that either (1) is a significant improvement over existing products or (2) performs a new function for the consumer. New products (which may be new to the company, if not necessarily the marketplace) are the lifeblood of any company and of the free-market system.

To Stay Ahead of or Match the Competition

History is full of examples of companies that thought they were dominant in their fields and failed to recognize how important a competitor's development was.

This is evident with technology and how quickly a product can become obsolete. Take, for example, the advancements in smartphones. Every year companies continue to try to stay ahead or at least match the competition.

To Continue to Expand Revenues and Profits

Some small businesses may be content to have the same earnings every year. But other businesses need to grow to continue rewarding shareholders. And to do that, they need to introduce new products.

For example, a company that owns a quick service restaurant may want to open more stores so that it can build a franchise model and scale to become a regional or national brand.

To Fill Out a Product Line

Some companies may offer certain products and it makes sense for them to create new products that are a natural extension of what they offer to fill out a product line.

For example if a company made dog food, it would make sense for it to launch treats and canned food.

To Take Advantage of an Opportunity

For a period of time, there may be an opening in the market and companies may want to capitalize on a trend.

For example, there may be information that comes out about a new "super food" and companies may take advantage of the trend by creating products that feature that food or ingredient.

Decisions

Mark, did you get a chance to look at those concession stand sales reports I sent you?

Yes, I did, and it sure looks like we are selling a lot of soft drinks but not so much beer, and beer is where we make the most money.

Good eye! I think we need to expand our beer selection to give customers some options and this would bolster our _____.

- **A.** product line
- **B.** beer expenditures
- **C.** investor expectations

Yes, and this will directly impact our _____.

- **A.** marketing mix
- **B.** revenues and profits
- **C.** competition

Correct Answers: A; B

Understanding the 4 Ps

Once a firm has determined that it needs a new product, the next challenges are:

- To conduct research to determine opportunities and challenges
- To identify the target market
- To determine the strategies for the marketing mix

The Process of Building a Marketing Mix

A **marketing mix** consists of the four key strategy considerations called the 4 Ps: product, pricing, place, and promotion strategies. Specifically, the marketing mix involves:

1. Developing a product that will fill consumer wants

2. Pricing the product

3. Distributing the product to a place where consumers will buy it

4. Promoting the product

All these blended together constitute a marketing program.

Let's see how the marketing process works.

Conducting Research and Determining the Target Market

The marketing process begins with conducting a survey or research to determine whether there's a market for the product the company is considering producing. That research should help to establish the target market for the product.

The Product Strategy

A marketing program starts with designing and developing a product—a good, service, or idea intended to satisfy consumer wants and needs. The designers must consider shape, size, color, brand name, packaging, and product image.

In reaching these decisions, a company needs to consider such matters as how well the product differs from other products. It also may do concept testing and test marketing to get a sense of consumer likes and dislikes.

- **Concept testing** is marketing research designed to solicit initial consumer reaction to new product ideas. That is, you might go out among the population you've identified as your target population and ask them if they think your potential product is a good idea.

- **Test marketing** is the process of testing products among potential users. That is, you try out a sample of the potential product among the target population to see what they think of it.

The Pricing Strategy

Pricing is figuring out how much to charge for a product—the price, or exchange value, for a good or service. The price of a product can depend on whether you have competitors, whether you need to offer low prices to get customers in the door, and the like.

The Place Strategy

Placing, or distribution, is the process of moving goods or services from the seller to prospective buyers. For example, retailers have expanded their abilities to serve customers who want products delivered to them directly by enhancing their e-commerce sites.

The Promotion Strategy

Promotion consists of all the techniques companies use to motivate consumers to buy their products—techniques such as advertising, public relations, publicity, personal selling, and other kinds of sales.

Decisions

Mark, you've picked up this information really quickly. I know Jim will be excited, and I'm sure he'll want to meet with you to discuss your ideas about potential new products.

Okay, but I'm a little nervous.

My guess is that Jim will probably ask you certain questions, based on the products that we mentioned earlier—primarily ticket options. It'll be really helpful if you have some of your responses ready.

A little bit later, Jim approaches Mark to chat.

Hi Mark! Ginger said you have ideas for getting more ticket sales?

Yes, as I see it, the primary product of the Doods is ticket sales for baseball games, but customers do not feel the _____ relates to value for them.

A. placement

B. promotion

C. price

D. product

Interesting. So if our single ticket price was lowered by $2 or so on week night games, do you think this would help get more customers?

Possibly so, but some of the problem is many of the customers said they don't really even know when the home games are, so I think we need to couple this price drop with more _____.

A. placement

B. market research

C. promotion

D. price drops

You know, Mark, you are asking for a price drop and more expenditures at the same time. That is going to cost quite a bit, so are you sure we will make enough money to cover this?

I think we will because I have spent the last six weeks
_____ this idea with customers, and a large percentage
of them agreed this would be more valuable to them.

A. concept testing

B. placing

C. marketing

D. market mixing

Let's hope so! Everything is a bit of a gamble, but
your research helps greatly!

Market Research

By the end of this lesson, you will be able to:

- Differentiate among the steps of the marketing research process.
- Differentiate among forces in the external marketing environment that can affect marketing strategy.

The Power of Knowing Your Market

"Making Informed Marketing Decisions with Research."

Mark has proven himself to be a great intern. He is not only learning a great deal about marketing but also contributing to the team with his creativity. As a result, Jim and Ginger want to involve Mark in gathering feedback from customers through market research.

Mark, I really like the innovative ideas you are coming up with for us!

Yes, I think we can really utilize your talent and work ethic to help us assess some marketing data from last year as well as build our market research plans for this season.

Let's go over the role market research plays in marketing. Then we'll give you a chance to help us better understand our market.

I'll get you started!

In this lesson, we'll help Mark understand the Doods' market by learning more about market research and the marketing environment.

Understanding the Marketing Research Process

Marketing research, part of the process of determining the 4 P marketing mix, is a four-step process of gathering and analyzing data about problems relating to marketing products, aiming to provide accurate information to marketers. Besides the marketing mix, marketing strategy must take into account the external marketing environment, which consists of seven outside forces.

Marketing Research: Getting Accurate Information to Make Marketing Decisions

For a marketing program to be successful, it depends on something crucial: accurate information. Accurate information is the province of **marketing research**, the systematic gathering and analyzing of data about problems relating to the marketing of goods and services.

Among other things, marketing research can tell you what consumers think about your firm's products, how satisfied they are with them compared with competitors', the effectiveness of your ads, what the sales potential is of new products, and what price changes might do to sales.

The marketing research process consists of four basic steps.

Define the Problem: Clarify the Question to Be Answered

Marketing professionals start defining the problem by analyzing the following questions:

- What is the present problem?
- What are the opportunities?
- What information is needed?
- How should we collect and analyze data?

Collect Facts: Use Published Data or Interviews, Observation, Experimentation, and Focus Groups to Get Information

Marketing research draws upon two kinds of data—secondary and primary. Most market researchers start with secondary data because it's cheaper and easier, although it has some disadvantages.

Secondary data is information acquired and published by others. Examples: U.S. Census Bureau data, various government publications, newspapers, magazines, academic journals, Internet searches, and blogs are all examples of secondary data sources.

Primary data is data derived from original research, such as that which you might conduct yourself. Examples: Direct observation, interviews, surveys, questionnaires, customer comment cards, and concept testing are all different sources of primary data. Some other important sources:

- **Focus groups** are small groups of people who meet with a discussion leader and give their opinions about a product or other matters. For example, a car company may have a group of consumers test drive a new model car and then give there feedback on the driving experience and features of the automobile.

- **Databases** are integrated collections of data stored in computer systems. In big companies, databases can be huge—so-called data warehouses—and allow market researchers to perform data mining, do computer searches of the data to detect patterns and relationships, such as customer buying patterns.

- **Neuromarketing** is the study of how people's brains respond to advertising and other brand-related messages by scientifically monitoring brainwave activity, eye tracking, and skin response.

Analyze the Data: Use Statistical Tools to Determine the Facts

Once data has been gathered, marketing researchers need to consider whether it needs to be treated further to make it useful. It may need **editing**, or checking over to eliminate mistakes. It may require the application of **data analysis**, subjected to statistical tools to determine its significance.

Take Action: Implement the Best Solution

Finally, with all the data and analysis in hand, the decision makers must decide how to use it—to determine the best solution and how it should be implemented.

Decisions

Ginger, I have been giving our product line a lot of thought, and I am drawing a blank on how to market our Dog House product, let alone come up with new products to sell.

I know how you feel! I really struggled with trying to figure out what our customers would want at the game when I first started, so I ended up asking a bunch of people at the games to get their feedback.

Yes! That is _____, and that is probably what I should do.

    A. secondary data

    B. the marketing process

    C. market research

I found it to be a great help, but I really needed more time with these people to ask them even more questions about what they like and don't like. I should have _____.

    A. collected secondary data from this group

    B. formed a focus group

    C. started the marketing process

Now you're talking! That would have been a great way to get the information we need directly from our customers. Who would have guessed that I'd actually be applying the concept of _____ in action this summer? Wow!

    A. primary research

    B. secondary research

    C. customer relationship processing

Correct Answers: C; B; A

Understanding the Marketing Environment

In addition to developing a marketing strategy that involves (1) identifying the target market and (2) determining the right marketing mix, you must also begin (3) dealing with the external environment—specifically the external **marketing environment**, the outside forces that can influence the success of marketing programs. These forces are (1) global, (2) economic, (3) sociocultural, (4) technological, (5) competitive, (6) political, and (7) legal and regulatory.

The Marketing Environment: Outside Factors That Influence Marketing Programs

To understand this environment, marketing managers working on marketing strategy need to do **environmental scanning**—looking at the wider world around them and identifying what factors can affect the marketing program.

Marketers usually can't control the external environment, but they need to understand how they are hindered or aided by it.

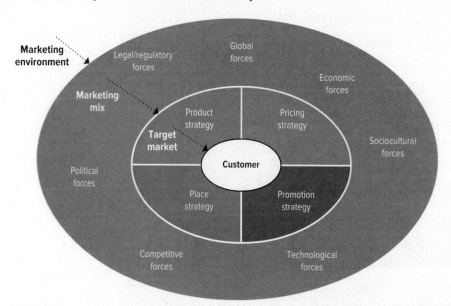

Global Forces

Global forces consist of influences brought about by all our global interconnections. For example, sites like Alibaba have increased access to foreign suppliers for individuals to be able to launch new businesses and source products in different markets.

Economic Forces

Economic forces—recessions, inflation, and the like—certainly affect consumers' buying power and willingness to buy.

For example, the Great Recession deeply impacted the housing market through foreclosures. Through the recovery process, interest rates remained low for a considerable amount of time to allow home buyers to lock in historically low interest rates.

Sociocultural Forces

Sociocultural forces include cultural changes reflecting customs, beliefs, and lifestyles of groups that differ in social class, ethnicity, age, and so on.

For example, companies may adjust their marketing efforts to become more diverse and inclusive to better reflect their customers.

Technological Forces

Technological forces consist of influences both highly visible that affect retailing, such as being able to quickly build online e-commerce sites to launch products and companies, and less visible that affect manufacturing and distribution such as automation, which change the nature of delivering goods and services.

Competitive Forces

Competitive forces consist of the actions of competing firms, industries, or countries. For example, as consumers begin consuming entertainment through devices, more companies are developing services and business models to compete for these consumers.

Political Forces

Political forces are influences that occur because of the decisions of politicians and public officials. Based on legislation that is enacted and policy decisions at the local, state, and federal level, companies may need to adapt to stay viable. Companies need to also look for new business opportunities as a result of political forces.

Legal and Regulatory Forces

Legal and regulatory forces consist of laws and government regulations designed to protect consumers and restrain anti-competitive business behavior. Companies must also stay aware to stay in compliance with laws.

Decisions

Mark has learned a great deal about marketing through his internship and is ready to help Jim and Ginger have a successful season marketing the Doods!

Ginger, I spent last night _____ of the Doods in order to narrow down what forces are impacting us the most right now, and I think it's the economy.

   A. scanning the environment

   B. researching the business

   C. watching the game on television

   D. planning marketing research

Really? How do you know?

Because I have tracked the ticket sales of the Doods over the last 20 years, and it seems to follow directly with the economy. When the economy is doing well, then the Doods sell more tickets regardless of how the team is doing.

Excellent analysis! Jim and I have been stuck on the idea that the city just goes in cycles and falls in and out of love with the Doods based on how they were playing that season. We thought ticket sales were driven by _____ forces.

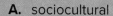

A. sociocultural

B. political

C. economic

D. competitive

That's right, but maybe you are on to something, Mark! If this is true, and we are experiencing a strong surge in the economy, then we might want to try to _____.

A. raise ticket prices right now

B. lower ticket prices to accommodate the situation right now

C. scan the political environment for opportunities

D. cut back on the sale of beer right now

I think so! This really might turn out to be a winning situation for the Doods!

Correct Answers: A; A; A

Introduction to Marketing: Test

1. Once Ned and Brad realized they had a problem with their leaky windows, they asked friends and looked on the Internet to see what options they had to solve the problem. This represents which step of the consumer buying process?

    A. problem recognition

    B. information search

    C. evaluating alternatives

    D. purchase decision

    E. postpurchase evaluation

2. Luz Maria and Mohammed have different ideas about what color to paint the house. Luz Maria wants a bright blue like her childhood bedroom in Mexico, and Mohammed wants a steel gray like his mosque's prayer room. Which element of consumer buying behavior is most likely influencing their decisions?

    A. culture and subculture

    B. social class

    C. reference groups

    D. personal image

    E. situational matters

3. An upscale jewelry store was considering moving into a transitional neighborhood with millennials in new housing mixed with older homes and lower incomes. Which customer attribute is most likely to be an issue?

    A. Customers have access to the product.

    B. Customers have decision-making power to buy the product.

    C. Customers have money to afford the product.

    D. Customers are not already served by competitors.

    E. Customers have a particular need that the firm can serve.

4. _____ segmentation consists of categorizing consumers according to statistical characteristics of a population, such as gender, age, income, and so on.

    A. Demographic

    B. Geographic

    C. Psychographic

    D. Benefit

    E. User rate

5. Beach Shack tends to locate its stores in tourist areas in beach towns. Which type of marketing segmentation does Beach Shack use?

    A. demographic

    B. geographic

    C. psychographic

    D. benefit

    E. physical

6. Action Vacations promotes trips to risk-oriented individuals who enjoy physical activity, like mountain hiking, helicopter skiing, ocean sailing, and white-water rafting. Action Vacations' market is most likely segmented by _____ segmentation.

    A. demographic

    B. geographic

    C. psychographic

    D. social class

    E. user rate

7. A commercial laundry targets restaurants and hotels that require uniforms for their staff. This is _____ segmentation.

    A. niche

    B. geographic

    C. customer-based

    D. B2C

    E. product-use-based

8. Which of the four Ps involves distributing the product to a place where consumers will buy it?

   A. product

   B. price

   C. place

   D. promotion

   E. public relations

9. Connie's Chalupas has placed samples of its product in grocery stores in mostly Hispanic neighborhoods to see how customers react to them. This company is conducting

   A. test marketing.

   B. competitive comparison.

   C. one-on-one marketing.

   D. concept testing.

   E. advertising testing.

10. Efron is in the stage of marketing research in which he needs to determine how he should collect and analyze data. This is the _____ step.

   A. define the problem

   B. collect facts

   C. analyze the data

   D. take action

   E. access databases

11. The United Way has a team that is responsible for creating content targeted at potential customers and clients in the community. This team is most likely a(n) _____ team.

   A. human resources

   B. marketing

   C. administration

   D. information technology

   E. accounting

12. Heritage Crest Bakery started out with a sourdough bread recipe. Over the years, it has added other styles of bread like rye, wheat, white, and potato. Together, the Heritage Crest breads can best be described as a

   A. product mix.

   B. promotional line.

   C. purchase point.

   D. product line.

   E. price segment.

13. When Apple releases a new iPhone, it often releases a captivating advertising campaign to bring awareness to the product. This statement best demonstrates which factor that influences buying behavior?

   A. situational matters

   B. culture and subculture

   C. social class

   D. personal image

   E. reference groups

14. August buys a puppy and takes her to his parents' anniversary dinner. August's sister Luz absolutely loves the puppy, and August encourages her to buy another puppy from the litter. After the party, Luz goes online and buys a puppy of her own. Which of the factors that influence buying behavior does this most likely represent?

   A. culture and subculture

   B. one-to-one marketing

   C. psychographic influence

   D. social class

   E. reference groups

15. Max Clubber is a famous retired boxer and stays in great shape due to the Superflex Home Gym system. Superflex hires him to be its spokesperson knowing that many individuals will buy the gym system because they want to look like Max. Superflex believes that Max will most likely impact which factor that influences consumer buying behavior?

   A. economics and pricing

   B. niche promotion

   C. personal image

   D. reference groups

   E. social class

16. Laguna Steelworks has divided its customer base into two groups—commercial and personal. It utilizes different pricing and promotion models for each group. Which concept best describes the process Laguna used to classify its customer bases?

   A. market segmentation

   B. niche marketing

   C. reference groups

   D. neuromarketing

   E. one-to-one attributes

17. Juniper Concrete is growing, and its management team has decided to create a formal marketing strategy. How should the company begin?

   A. by researching external marketing forces

   B. by identifying its target market

   C. by analyzing secondary data

   D. by preparing a marketing mix summary

   E. by enhancing its product line

18. Clarion Shampoo creates a new shampoo that is supposed to eliminate dandruff while also eliminating grey hair. It begins soliciting volunteers willing to have their hair professionally cleaned with the new shampoo. The volunteers are then asked for their feedback on the shampoo. This process can best be described as _____, a part of the _____ strategy step in the marketing process.

   A. concept testing; promotion

   B. test marketing; product

   C. concept testing; target market

   D. test marketing; research

   E. product testing; pricing

19. Revival Mineral Water has hired a web development company to create a new storefront app for mobile devices. It hopes that the storefront app will make it easier for customers to purchase its water in the future. This scenario would most likely be part of the _____ strategy in the marketing process.

   A. product

   B. user-to-user

   C. technology

   D. place

   E. promotion

20. The leadership team at Simplicity Uniforms recently hired a compliance officer who will be responsible for ensuring that the company is compliant with all relevant finance and accounting laws. Which external marketing force is most likely at play in this example?

   A. political

   B. global

   C. legal and regulatory forces

   D. ethical

   E. culture and subculture

6 Understanding Your Customer

What to Expect

When you hear about research, what is the first thing you think of? For many, its lab coats and clipboards. For entrepreneurs, marketing, research is used to understand customers – even you! Successful entrepreneurs are often good marketers and know how to interpret market research. Keep reading to learn what marketing research truly means.

Chapter Topics:

- **6-1** Marketing Research
- **6-2** Consumer Behavior
- **6-3** Business-to-Business Marketing

Copyright © McGraw Hill   Marco VDM/E+/Getty Images

Lesson 6-1
Marketing Research

This lesson introduces you to marketing research, not only as a marketer *doing* research but as a consumer *being* researched. How would you feel about being experimented on? This question raises common ethical concerns with marketing research, and this lesson also discusses the ethics of marketing research in an increasingly digital world.

By the end of this lesson you will be able to

- Explain the importance of marketing research to a firm.
- Outline the steps of the marketing research process.
- Identify the different types of marketing research data.
- Describe the role of experiments in marketing research.
- Explain why the need to understand competitors as well as customers is an Important part of marketing research.
- Describe the main ethical issues in conducting marketing research.
- Explain the difference between marketing research and marketing analytics.

Marketing Analytics Implications

- Marketing research and marketing analytics work hand in hand, but they are not the same thing. Marketing research provides data to answer a specifically defined question. Marketing analytics explores the data to gain deeper insights about the data that collection alone cannot provide.
- Marketing analytics helps firms identify purchasing patterns; data mining is one method that firms use to explore and interpret data.
- Marketing analytics helps firms identify and interpret the relationships among multiple, complex sets of data; data modeling has become one of the fastest-growing methods for this type of data analysis.

LEGO: Building New Customers through Marketing Research

Believe it or not, marketing research has completely transformed countless companies, including LEGO. The LEGO philosophy is built on "systematic creativity—Lego bricks are all part of the Lego system, which essentially means that they can easily be combined in innumerable ways—and just as easily be dismantled. The more Lego bricks you have, the more fertile your creativity can become. The combination of a structured system, logic, and unlimited creativity encourages the child to learn through play in a wholly unique Lego fashion."

After reading this philosophy, you'll most likely think it can be applied to both boys and girls (and it can!), but for many years, LEGO catered to only one market segment: young boys. Girls comprised only 9 percent of LEGO's users. This provided the company an opportunity to increase sales by expanding its reach and empowering young girls in the process. But how would it appeal to this female audience? By conducting marketing research, of course!

LEGO used marketing research to understand how consumers interact with their products. As a result of marketing research, the company launched the LEGO Friends line focused on young girls.

LEGO conducted primary research in the form of focus groups, interviews, and observation to study the play habits of 4,500 girls and their mothers globally. Researchers discovered that boys and girls tended to play differently. For example, after boys finished building a castle, they tended to grab other item such as figurines and swords and use the castle simply as a backdrop for a battle. Girls, on the other hand, tended to want to play with the castle once it was built, and quickly discovered there was nothing to do with the castle because the inside was rather plain.

With these findings, LEGO launched the "LEGO Friends" line focused on young girls. The new line included a cupcake café, a giant treehouse, a supermarket, and other construction sets. The sets contained just as many pieces as other sets but offered girls the ability to assemble, play, and then continue assembling. While there were critics of the girl-focused line, it proved to be a success for the company. LEGO has seen a 15 percent average growth in sales since its launch.

Now that LEGO was focusing on both boys and girls, there was still one market that was underserved by LEGO: adults. Once again, LEGO conducted marketing research and discovered that adult LEGO users spend significantly more per year than average child users. One user even reported spending $50,000 in one year! This new information led the company to decide to sell more expensive sets, like the Star Wars Millennium Falcon, priced at $800 with 7,541 pieces.

The Importance of Marketing Research

Marketing research is the act of collecting, interpreting, and reporting information concerning a clearly defined marketing problem. When done properly, marketing research helps companies understand and satisfy the needs and wants of customers.

For example, LEGO's marketing research is aimed directly at its customers. LEGO:

- Collects information on its market segments by observing how children play and interact with the product.
- Interprets information by outlining the play process from start to finish for boys and girls.
- Reports its results by recommending the development of new products that match different types of play.

Marketing research has become more important, and more complicated, as markets continue to become globalized and product life cycles become shorter. In such a climate, companies need accurate information to reduce risk and make good decisions. This quest for information compels companies to spend billions of dollars each year on marketing research.

Patterns of consumer behavior that can change quickly, such as trends, styles, and preferences, also provide an incentive for firms to acquire fresh information through marketing research. Firms can no longer rely only on historical data to determine future marketing strategies. Now, they must generate timely information, interpret it quickly, and take action before the competition does. If they don't, they will be beaten before they can even begin.

Marketing research impacts almost every aspect of a company's business. We can see this impact clearly in terms of the four Ps: product, price, place, and promotion.

Product

Products need to be developed based on real customer needs and wants, not just the whims of marketing departments. Product developers in research and development departments must have an idea of what customers want before they can create new products, and finance departments must have good information concerning the viability of a product to approve expenditures on new product development.

Price

Pricing requires analysis of the size of the potential market and the effects of price changes on demand. This information comes from **demand analysis,** a type of research used to estimate how much customer demand there is for a particular product and to understand the factors driving that demand.

Place

Decisions regarding the place or distribution function must be made using **sales forecasting,** which is a form of research that estimates how much of a product will sell over a given period of time. Using this research, firms know how much product to hold in inventory at various points in the distribution network. We discuss sales forecasting in more depth later in the lesson.

Promotion

Promotional activities such as advertising must be evaluated based on their effectiveness. Firms use advertising effectiveness studies and sales tracking to gauge how well advertising and promotional campaigns are working. **Advertising**

effectiveness studies measure how well an advertising campaign meets marketing objectives such as increasing market share, generating consumer awareness of a product, or creating a favorable impression of the company's products. **Sales tracking** follows changes in sales during and after promotional programs to see how the marketing efforts affected the company's sales. Marketing professionals use the results of this research to adjust where and how they apply promotional efforts.

In addition to its impact on firms, marketing research matters to consumers like you. Consumers rely on companies to develop and market the products they need and want. Without marketing research, companies would mostly be guessing.

The Marketing Research Process

Marketing researchers use a variety of research techniques, and gather data from an assortment of sources, to provide the goods and services people truly desire. Regardless of the technique employed, firms follow five basic steps when they engage in the marketing research process.

STEP 1	STEP 2	STEP 3	STEP 4	STEP 5
Problem definition	Plan development	Data collection	Data analysis	Taking action

Step 1: Problem Definition

Problem definition is the first step in the marketing research process. Often firms know that they have a problem but cannot precisely pinpoint or clearly define the problem. Clarifying the exact nature of the problem prevents the firm from wasting time, money, and human resources chasing the wrong data and coming up with the wrong solutions.

To begin, a firm should set specific research objectives. As with overall marketing objectives, research objectives should be specific and measurable. They represent what the firm seeks to gain by conducting the research.

For example, LEGO started by defining a problem such as "Why are girls not interested in our products?" The genesis of any company comes from finding a better way to serve customer needs. For LEGO, focus on the initial problem definition guided the research process and ultimately led to a viable product solution. In other words, without clearly defining the problem, LEGO would have never found that girls tend to play with the product differently than boys and that they needed products to match their type of play to keep them interested.

Step 2: Plan Development

Plan development, often called *research design,* involves coming up with a plan for answering the research question or solving the research problem identified in the first step of the process. Plan development is about identifying what specific type of research will be used and what sampling methods will be employed.

Types of Research

How marketing researchers decide which type of research they need to do depends on the nature of the question or problem the firm has and how well the researchers understand it. There are three general types of marketing research:

1. **Exploratory research** is used when researchers need to *explore* something about which they do not have much information. Exploratory research seeks to discover new insights that will help the firm better understand the problem or consumer thoughts, needs, and behavior. This type of research usually involves a personal interaction between the researcher and the people being researched, often in the form of conversations, interviews, or observations.

2. **Descriptive research** is used when researchers have a general understanding of a problem or phenomena but need to *describe* it in greater detail in order to enable decision making. Descriptive research seeks to understand consumer behavior by answering the questions who, what, when, where, and how. (Note that descriptive research *cannot* answer "why" questions.) Examples of descriptive information include a consumer's attitude toward a product or company; a consumer's plans for purchasing a product; specific ways that consumers behave, such as whether they prefer to shop in person or online; and demographic information such as age, gender, and place of residence.

3. **Causal research** is used to understand the *cause-and-effect* relationships among variables. Causal research, also called *experimental research,* investigates how independent variables (the cause) impact a particular dependent variable (the effect).

Almost all causal research and most descriptive research starts with the development of a hypothesis. A **hypothesis** is an educated guess based on previous knowledge or research about the cause of the problem under investigation. For example, based on your life experience, you could *hypothesize* that people are more likely to wear shorts in the summer than in the winter. Although this is almost certainly true, we cannot actually prove that it is true without collecting evidence and using facts to empirically support our educated guess.

Marketers will establish specific data collection procedures, discussed in the next section, based on the general type of research the company decides is most appropriate to achieve the research objective. Companies need not limit themselves to one type of research and often use multiple approaches to help solve a problem. In many cases, marketers will begin with exploratory methods to gain a better-nuanced understanding of a marketplace phenomenon in order to make hypotheses about consumer behaviors, and then later use experimental methods to test whether their hypotheses are true.

Summary of Different Types of Research

Research Type	When Used	How Conducted	Type of Hypotheses
Exploratory	Typical when information is limited, such as when a firm enters a new market	Interviews and/or observation	Questions designed to gain broad understanding
Descriptive	For situations where specifics of a market are not well defined (i.e., who, what, when, where, how)	Surveys and/or focus groups	Multiple and specific questions to gain specific understanding
Causal	Used in situations where clarifying what caused an action to happen, such as "why are sales increasing at only some stores?"	Experiments, often in a store setting	Questions that assess why something happens

Sampling

It would be impossible—from both a budget and time perspective—for marketing professionals to obtain feedback from all the members of their target market. Instead, they must rely on sampling. **Sampling** is the process of selecting a subset of the population that is representative of the whole target population. Feedback gathered from this sample can then be generalized back to the entire target market. How researchers conduct sampling is critical to the **validity** of the research findings—how well the data measure what the researcher intended them to measure.

Sampling can be broken down into two basic types:

- **Probability sampling** ensures that every person in the target population has a chance of being selected, and the probability of each person being selected is known. The most common example of probability sampling is **simple random sampling,** where everyone in the target population has an equal chance of being selected. Simple random sampling is the equivalent of drawing names from a hat; every name in the hat has an equal chance of being chosen.

- **Nonprobability sampling,** on the other hand, does not attempt to ensure that every member of the target population has a chance of being selected. Nonprobability sampling contains an element of judgment in which the researcher narrows the target population by some criteria before selecting participants. Examples include **quota sampling,** in which the firm chooses a certain number of participants based on selection criteria such as demographics (e.g., race, age, or gender), and **snowball sampling,** in which a firm selects participants based on the referral of other participants who know they have some knowledge of the subject in question.

While probability sampling enables researchers to generalize findings from a portion of a target population, nonprobability sampling can generate findings that may be more appropriate to the research question.

Sampling is a valuable way to gather feedback about a large target market when obtaining feedback from every member of that target market is not feasible. In this photo, an Earth Balance representative is handing out free soy milk samples to consumers at a local market, and then asking them to complete a quick product survey—the link for which is found on the postcards displayed on the table.

Step 3: Data Collection

The third step of the marketing research process, data collection, begins with a decision about data. Researchers must ask themselves: Can we acquire the data we need to answer our questions from someone else, or do we need to collect the data ourselves? This section focuses on methods used when researchers must go out and collect the data themselves, and a later section discusses the advantages and disadvantages of using data already collected by someone else.

Primary data collection is when researchers collect data specifically for the research problem at hand; this can be either qualitative or quantitative in nature. Qualitative research studies the *qualities* of things, whereas quantitative research examines the *quantities* of things. **Qualitative research** is characterized by in-depth, open-ended examination of a small sample size, like in-depth interviews or focus groups, and is used for exploratory and descriptive research. **Quantitative research** is characterized by asking a smaller number of specific and measurable questions to a significantly larger sample size and is used for descriptive or causal research. The figure below shows the relationships among research objectives and the collection methods and types.

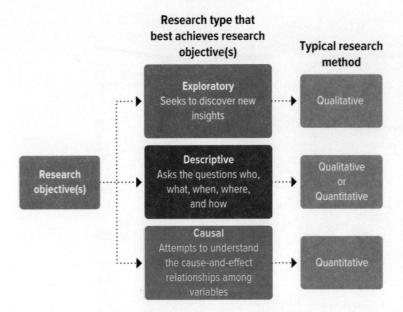

The Relationships among Research Objectives and Collection Options

Qualitative Research

As noted above, qualitative research includes exploratory types of research such as interviews, focus groups, and observation:

- In an **interview,** the researcher typically works with one participant at a time. During the interview, the researcher asks open-ended questions about how the individual perceives, uses, or feels about various products or brands. Interviews can be useful in figuring out what or how people think, but they have the limitation of being very time-consuming and must be conducted by experienced researchers who know how to properly ask and follow up on questions in a way that elicits rich and useful information from participants.

- **Focus groups** are conducted by a moderator and involve interviewing a small number of people (usually 8–12) at a time, where participants interact with one another in a spontaneous way as they discuss a particular topic or concept. The interactive nature of this setting lends itself to drawing out opinions and generating insights into a marketing question, much like one-on-one interviews but with the added advantage of group interactions. However, focus groups are expensive to perform and must be led by experienced moderators that are able to direct group discussions and manage interpersonal dynamics.

Qualitative methods like interviews and focus groups can provide researchers a great deal of insight, but they don't always allow researchers to draw generalized conclusions about the larger consumer population. To collect the necessary data to achieve their research objective, companies often turn to quantitative research.

Quantitative Research

Quantitative research includes surveys, experiments, and mathematical modeling:

- **Surveys** or questionnaires pose a sequence of questions to **respondents.** They provide a time-tested method for obtaining answers to who, what, where, why, and how types of questions and can be used to collect a wide variety of data. In addition, they help determine consumer attitudes, intended behavior, and the motivations behind behavior. Surveys often employ multiple-choice questions, making them appropriate for gathering feedback from a large number of participants. They can be administered by mail, at shopping malls, on the telephone, or online.

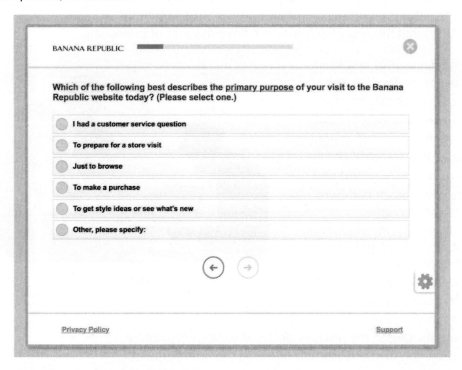

Banana Republic conducts survey research of its customers to find out many things, including their primary purpose for visiting its website.

- **Experiments** are procedures undertaken to test a hypothesis. They allow researchers to control the research setting so that they can examine causal relationships among variables. Researchers can see how a change in an independent variable (for example, price or product package color) might cause changes in another dependent variable (for example, sales or customer preferences). Experiments traditionally take place in a simulated environment known as a laboratory, but researchers can also perform **field experiments** in natural settings like stores or malls. While field settings offer an element of realism, they are also more difficult to control, which can lessen the validity of the experiment.

- Another type of causal research involves **mathematical modeling,** in which equations are used to model the relationships among variables. Statisticians have developed fairly reliable models for predicting certain elements of market behavior under normal market conditions, like how much consumer demand for a given item would go up or down based on a price change.

Step 4: Data Analysis

The purpose of data analysis is to convert the data collected in Step 3 into information the firm can use to answer the question or solve the marketing problem originally identified. If a research hypothesis was developed to be tested, analysis of the data should allow marketers to accept or reject the hypothesis.

Analyzing Qualitative Data

Researchers can gain substantial insights from qualitative research. Qualitative data gathering and analysis can give researchers ideas about the subject that can later be tested through quantitative research. Because there are no predetermined sets of responses (e.g., multiple-choice answers), the participant can open up to the interviewer and cover things that the researcher may not have thought of previously.

Because qualitative research usually results in great quantities of textual or media data, the qualitative analysis process generally involves a systematic approach to summarizing the data called coding. **Coding** is the process of assigning a word, phrase, or number to a selected portion of data so that it can later be easily sorted and summarized. The words, phrases, or numbers assigned to the data are called *codes*.

NVIVO software is used by researchers to sort through qualitative data and assign codes that help analyze the data.

Data analysis, particularly when applied to qualitative data, can be challenging. Due to the immediate and personal involvement of both the researcher and research participants in the process, results may be difficult to measure objectively and without bias. Qualitative data, due to the open-ended and exploratory nature, often requires the interpretation of subtleties and nuances that inexperienced researchers may miss.

Analyzing Quantitative Data

Quantitative methods for collecting and analyzing data can often be done quickly, at relatively low cost, and can help researchers describe large groups of customers and activities and understand cause-and-effect patterns in consumer behaviors.

Quantitative analysis almost always involves the use of **statistical analysis,** which is the mathematical classification, organization, presentation, and interpretation of numerical data. Simple analysis can be done in common programs like Microsoft Excel, whereas more advanced statistical modeling requires expensive and dedicated software like IBM's SPSS. There are two main types of statistics: descriptive and inferential.

Descriptive statistics are used to *describe* characteristics of the research data and study sample. Descriptive statistics, called "descriptives" for short, speak only to the properties of the data on hand and cannot be used to make claims about the larger population from which the sample was drawn.

Descriptives are the most common type of statistics we see in our daily lives, and they are the most straightforward way to tell a story about data. Anytime you see a percentage, a ratio, a bar graph, or a pie chart, that's a descriptive statistic. Descriptives can detail one variable at a time or multiple variables at a time. For example, your GPA uses one number to tell a story about your academic performance overall, across the dozens of classes you've taken. But statisticians could dig a bit deeper to introduce a second variable and compare your GPA for required versus elective courses.

Keep in mind that although descriptives allow us to summarize large quantities of data with a small amount of information, just like any type of summary, we lose a bit of detail along the way. For example, by looking at a GPA of 3.46 we would have no way of knowing whether a given student consistently earned B+ grades in all of his classes, or whether she was almost a straight A student that received a single D last semester.

Inferential statistics, on the other hand, are used to make *inferences* about a large group of people from a smaller sample. The purpose of inferential statistics is to make generalizable conclusions about a population by studying a small group from that population. Inferential statistics almost always include two or more variables.

Some inferential statistics are used to explore *differences* between two groups of people. For example, we might want to explore the differences between how men and women perform on an exam (studying two groups of people at the same time), or see if there is a difference in exam performance between the midterm and the final (studying the same group of people at two different times).

Other types of inferential statistics examine *relationships* among variables. For example, we might want to see if there was any correlation between time spent studying and exam grade performance.

As with qualitative data, there are drawbacks to quantitative analysis. While statistical analysis might give marketing professionals insight into consumer behavior, whether the results should be accepted depends on how well the sample represents the general population and how well the statistics were conducted. Remember that statistical results are only as good as the data you have, and the data are only as good as the methods you used to collect them and the sample you collected them from. As we mentioned earlier, validity concerns how well the data measure what the researcher intended them to measure. If you have ever come across a statistic in the newspaper or in an advertisement and thought "that doesn't seem right" or "yeah, but . . ." that's calling the validity of the results into question. Often, validity is called into question because of poorly worded questions that can be interpreted in several ways.

The following table lists some of the advantages and disadvantages of qualitative and quantitative research methods.

Advantages and Disadvantages of Qualitative and Quantitative Research

Research Method Type	Advantages	Disadvantages
Qualitative	• Uncovers details concerning the motivations behind behaviors • Is not limited to a predetermined set of responses • Can be a good way to start research into a marketing problem • Can be very flexible in approach • Can be used to generate marketing ideas	• Results may be difficult to measure objectively • Research can take longer than quantitative methods • Potential for researcher bias • Individual participants may not represent general target market • Small sample size
Quantitative	• Results may be generalizable to a larger population • Some methods can be conducted quickly and inexpensively • Analysis of data can be faster than in qualitative research • Can conduct causal studies that indicate why behaviors occur • Can be cost-effective • Often convenient for respondent	• May be limited by researchers' questions • Response rates can be very low • Difficult to determine nonresponse bias • Possible respondent self-selection bias • Participant resistance to giving sensitive information

Step 5: Taking Action

The culmination of the marketing research process is a formal, written report to decision makers. The report typically includes a summary of the findings and recommended actions to address the problem. Often, an oral report on the project is presented along with a written one.

Research report findings should be presented in a clear and understandable manner and include appropriate visual data, such as figures and tables, to support the findings and recommendations. The research report should allow the marketing manager to solve the marketing problem or provide answers to the marketing manager's questions.

Both reports should communicate any limitations of the research. Limitations could include a variety of things, including inadequate sample sizes or samples that do not adequately represent the population under study. It is important that marketers honestly discuss the limitations of the research. Such limitations should be considered before the firm makes any final decisions based on the research.

Step 1: Problem Definition

Redbox has a problem. Rental of DVDs at its self-serve kiosks have been steadily declining. This problem helped Redbox to define a problem and set the research objective: can Redbox compete in the online streaming market?

Step 2: Plan Development

To study this objective, Redbox was able to determine that its 27 million customers (enrolled in its loyalty program) would be a suitable test group and that it needed to study consumer preferences for online streaming.

Step 3: Data Collection

Redbox conducted primary research through surveys and focus groups comprised of current disc renters. These primary data were used to uncover unique needs and wants of Redbox customers.

Step 4: Data Analysis

The results of Redbox's research uncovered a market need for a non-subscription content streaming service that focuses on new release movies (more titles than competitors) and lower rental process (starting at $3.99) for a longer duration (2 days).

Step 5: Taking Action

The report of the research findings helped Redbox to design and launch its new streaming service. The new service focuses on the needs of 27 million customers and also hopes to attract new customers.

Marketing Research Data

Data are facts or measurements of things or events. *Qualitative data* might take the form of interview transcripts, video recordings of focus groups, or field notes from an in-store observation of shoppers. *Quantitative data* may be a collection of filled-out questionnaires or a spreadsheet with thousands of rows and columns of responses to an online survey. Data can also be presented in the form of a graph for easier digestion.

Data are the building blocks of research. Without data, researchers would simply be guessing when making marketing decisions about strategy, product design, pricing, distribution, promotions, and advertising. Likewise, without data, we cannot tell a story about our sample, but data by themselves do not tell the story. Researchers need to analyze data to help tell the story and turn the data into information.

Information is the result of formatting or structuring data to *explain* a given phenomenon, or to *define the relationship* between two or more variables. Information is the result of data analysis. Qualitative information from interviews may take the form of a theory about customer behavior or insights about why certain customers make the choices they do when they shop. Quantitative data can be presented using charts or diagrams and can provide information about customer behavior such as who a store's customers are, when they shop, what they buy, and how they move through the store. Unlike data, information can support and enable marketing decision making.

Survey engines such as Survey Analytics help researchers collect data through its survey platform; it can also automatically generate a bar or circle graph, like the ones shown here, out of the complex data it collects, making the data easier for marketers to digest and interpret.

Primary versus Secondary Data

As discussed previously, all research endeavors must begin with researchers asking themselves about the nature and type of data they need to answer their questions. Sometimes questions can be answered using existing data, but sometimes they cannot, and researchers must go out to collect new data to answer their particular question.

Primary data are data that are collected specifically for the research problem at hand. For example, LEGO conducted MRI scans of children's brains as they played to understand how their brains reacted. In doing so, the company collected primary data because the data were collected specifically for the purpose of this research.

Secondary data are data that have been collected for purposes other than answering the firm's particular research question. For example, if LEGO wanted to dig into its data to see what the top three selling products are for children under age 10, this would be a secondary use of the company's primary data because the data were collected for another purpose.

As with most decisions, using secondary data requires the marketing researcher to make a trade-off: On one hand, secondary data are much less expensive and easier to access than primary data, but on the other hand, secondary data collected for another purpose may not shed light on the specific nuances of a firm's particular problem.

Primary data collection may be necessary if secondary information cannot adequately answer the research question. Using primary data requires marketing researchers to invest far more time and money into their research endeavors, with respect to both collecting and analyzing data, but this can result in far more firm- or problem-specific information that allows for a more nuanced understanding of customers and their behavior.

(You're probably curious about the results of this research. The MRI scans showed that children had the highest level of engagement in skill-based play and from gaining the satisfaction of completing the building process. This research finding led LEGO to rank its sets based on skill level.)

Differences between Primary and Secondary Data

	Primary Data	Secondary Data
Collection Method Examples	• Focus groups • Surveys • Observations • Data gathered by equipment (e.g., video) • In-depth personal interviews	• Literature reviews • Online electronic searches • Company records • Marketing information systems • Private research companies • Boundary spanners (e.g., salespersons)
Advantages	• Pertain only to firm's research • May provide insight into why and how consumers make choices	• Less expensive (often free) • Information typically readily accessible
Disadvantages	• More expensive • May be difficult to enlist customer participation • May take excessive amount of time to collect	• Data may not be relevant • Data may not be accurate • Data may have been altered • Data may contain bias
Examples of Use	• To understand what motivates consumers • To determine the effect of variables (e.g., price) on product choice • To gain feedback on company's existing and proposed products	• To gather macroeconomic data • To gather socioeconomic data • To obtain information about competitors • To gain insight into international cultures and markets

Sources of Primary and Secondary Data

Primary data can be qualitative or quantitative in nature and can take the form of interviews, focus groups, observations, video recordings, questionnaires, surveys, or experiments.

Secondary data can come from internal or external sources. *Internal secondary data* are collected by the company and can include things such as sales by product, information about individual purchases from loyalty cards, previous research reports, accounting records, and market information from the sales force. Companies often build large internal databases in which to store such data.

External secondary data can come from many sources. Governments compile a lot of data and make them available to the general public. U.S. government agencies such as the Economics and Statistics Administration and the Census Bureau provide a great deal of useful secondary data in various publicly available reports. The Census Bureau, for example, provides geographic and demographic data about U.S. citizens for free. Other sources of secondary data, some free and some paid, include trade associations, academic journals, business periodicals, and commercial online databases. The Internet is a rich source of secondary information.

Sources of Primary and Secondary Data

Primary Data Sources	Secondary Data Sources
Observations	Previous research
Audio or video recordings	Literature reviews
Interviews	Online electronic searches
Correspondence	Company records
Focus groups	Marketing information systems
Case studies	Private research companies
Questionnaires	Government reports
Online surveys	Academic journals
Experiments	Periodicals and mass media
	Historical information

Experimental Research in Marketing Research

Earlier in this lesson, we briefly discussed marketing research experiments. As you will recall, experiments are procedures undertaken to test a hypothesis. Just like traditional scientists, marketing researchers are *social* scientists, and they conduct research experiments in much the same manner, and for much the same purpose.

Experiments are a type of causal research: the purpose is to allow researchers to control and then manipulate the research setting so that the researchers can examine causal relationships among variables. In a normal marketplace setting, there are lots of things going on at once, so it makes testing cause-and-effect relationships difficult. By carefully and intentionally manipulating—in the literal sense of skillful control and adjustment, not the colloquial sense of being unscrupulous—certain elements of the research setting, researchers are able to test cause-and-effect relationships in a much more definitive way.

Independent and Dependent Variables

Usually experimental research means investigating how a change in an **independent variable** might cause changes in one or more **dependent variables.** Dependent variables are the ones being tested and measured. A change in the dependent variable *depends* on something changing with another variable in the study. Independent variables, on the other hand, are the variables that *cause* the change in the dependent variables.

Some independent variables can be *controlled* for but not changed or manipulated specifically by researchers. Examples of this type of independent variable would be age, height, income, grade point average, weather, or time of day. But other types of independent variables can be intentionally *manipulated* by researchers, such as price, package color, store temperature, number of staff working, or product placement.

Research Considerations

In addition to considering how to study a problem (e.g., survey, experiment, etc.), marketers also need to determine who is being researched in different experimental conditions. **Experimental conditions** are the set of inputs (independent variables) presented to different groups of participants. In an experiment, typically only one or two inputs will be changed in each condition. This tight control over each condition allows researchers to identify what is causing something to happen in one group versus not happen in another. Those assigned to conditions are called **participants** because they are *participating* in the experimental conditions. Participants are typically selected using a defined selection process and randomly assigned to one of the different experimental conditions.

For example, we might be interested to research if the new low-calorie ice cream, Halo Top, contributes to weight loss. Halo Top is a low-sugar, high-protein ice cream that lists the calorie count for each pint prominently on the label. To test this question, we could set up three experimental conditions.

- **First Condition:** Participants eat only Halo Top ice cream (in specific quantities and with all other food intake being equal).
- **Second Condition:** Participants eat a higher-calorie ice cream (in the same specific quantities and with all other food intake being equal).
- **Control Condition:** Participants eat no ice cream (with all other food intake being equal).

This would be a very difficult experiment to conduct because it is hard to control exactly how much a person eats, especially outside of a laboratory. However, if this test could be conducted, we would expect to see differences in weight gain/loss between the groups. Marketers analyze these differences to determine, with all other things being equal, what happens when one thing is changed (in this hypothetical experiment, type of ice cream). The results of this experiment would seem obvious, but in many cases, determining exactly what is causing something to happen can be complex. In cases where it is important to test what is causing change (e.g., pharmaceutical trials, tire blowouts, etc.), experiments may be the only way to sort out a complex problem.

Experiments traditionally take place in a simulated environment: a laboratory—for social scientists as well as traditional scientists—because researchers can have more control over the research setting, environment, and extraneous variables. An **extraneous variable** is anything that could influence the results of an experiment that the researchers are not intentionally studying. For example, if one of our Halo Top participants were to eat a few extra protein bars each day during the experiment, the experiment's results would be skewed by the extraneous variable (protein bars).

For marketers studying consumer behavior, it often makes more sense, both practically and financially, to study customers in a realistic marketplace environment. As we discussed earlier, experiments where researchers control or manipulate variables but that are conducted in natural settings are called *field experiments.* Although field settings offer an element of realism, they are also more difficult to control and more extraneous variables can possibly influence the outcomes.

Deciding whether to conduct experiments in a laboratory or in the field is certainly a question of resources, but it is importantly a question of validity. As you'll recall, validity is the extent to which an experiment measures and tests what it is supposed to measure and test. There are two types of validity to consider, which highlight the trade-offs a researcher must make when choosing between laboratory and field experiments.

The first type considers the validity of the **experimental manipulation** itself. **Internal validity** is the extent to which changes in the outcome variable were actually caused by manipulations of the independent variable conditions. The higher the internal validity, the more faith we can have in the cause-and-effect relationship demonstrated. The second type of validity to consider is **external validity,** which is the extent to which the results of the experiment can be generalized beyond the study sample of subjects. The higher the external validity, the more likely it is that the cause-and-effect relationship demonstrated by the experiment will be found across similar real-world settings. Field experiments tend to be lower in internal validity but higher in external validity, whereas laboratory experiments tend to be higher in internal validity but lower in external validity.

For example, in the Halo Top study, if the experiment were conducted in a laboratory setting, it would be easier to control what the participants ate. Exact portions could be weighed and measured. Other variables could also be closely controlled, such as exercise and sleep, which might affect weight

loss. The problem is, not many people would want to live in a controlled lab setting for a long period of time.

Conversely, the study could be conducted as a field experiment. In this setting, participants would record their eating, exercise, and sleep. However, it is pretty easy to imagine that errors would affect the study results. The lab experiment is most effective, but unrealistic, and the field experiment is most realistic, but least effective. In most cases, there is no great solution, so these trade-offs are considered by marketing researchers and the best solution is followed, based on the objectives of the experiment.

In addition to considering how to study a problem, marketers also need to determine who to study in different experimental conditions. If all of these individuals participated in the Halo Top experiment in a laboratory setting, it would be easier to control what they ate and therefore observe the differences in weight gain/loss among them more accurately. The problem is, not many people want to live in a controlled lab environment for extended periods of time.

The Importance of Competitors and Consumers in Marketing Research

The purpose of marketing research is to help entrepreneurs and marketing managers make better decisions. Often when we talk about marketing research, we discuss it in the context of the consumer marketplace. Why people make the purchase decisions they do, and when, how, and where they make those decisions are common questions marketers ask. Entrepreneurs and marketeers also often want to understand what their customers want and how things such as product features and prices influence their decision making. When done properly, marketing research helps companies understand and then satisfy the needs and wants of customers, which creates value for all.

Another important component of marketing research involves gathering data about what competitors are doing in the marketplace. **Competitive intelligence** involves gathering data about what strategies direct and indirect competitors are pursuing in terms of new product development and the marketing mix. Such information can provide a firm with foreknowledge of a competitor's upcoming promotions or products, allowing it to respond in a way that blunts the effects of the competitor's actions.

A company can obtain information about another company's activities and plans in a number of ways, including conferences, trade shows, social media sites, competitors' suppliers, distributors, retailers, and competitors' customers. Competitor websites and the websites of government agencies such as the Securities and Exchange Commission (SEC) and the U.S. Patent and Trademark Office may contain financial and new product information about competitors. Firms also can obtain information for a fee from a number of companies that collect data and make them accessible.

There are also unethical ways of obtaining competitive intelligence. Bribing competitor employees for information and hiring people to tap phone lines or place surveillance cameras on the premises of rival companies are both examples of unethical and illegal ways of getting competitive information. Firms should scrupulously avoid these activities for legal reasons as well as for the damage they can do to a company's image and reputation.

Trade shows, such as the annual Consumer Electronics Show (CES) in Las Vegas, are a great place for marketers to gather competitive market intelligence.

Marketing Research Ethics

Whether conducting research for a domestic or international market, firms must always consider the ethical implications of gathering data on customers and competitors. Such considerations have become even more essential in a world of rapidly changing technology.

Privacy in a Digital Age

The increasingly powerful computer capabilities available to companies allow for the collection, storage, and analysis of data from millions of consumers. Online platforms like Twitter that are used by millions of consumers present firms with almost unlimited access to personal data. But companies must be careful not to go too far in collecting information of a sensitive nature. There is enough concern in the marketing industry about consumer privacy that many companies now have chief privacy officers who serve as watchdogs, guarding against unethical practices in their company's collection of consumer data.

At issue is the willful intrusion on the privacy of individuals. Consider mobile ad network companies that use unique identifiers embedded in smartphones to collect information about consumer preferences as they move from one app to another. To do this, such companies have worked around efforts by Apple to protect the privacy of iPhone and iPad users. The companies say they need personal data from users or they will lose millions of dollars of revenue from the firms that hire them, such as Mazda and Nike. Although Mazda and Nike don't participate in the actual tracking of data, they could potentially use what's collected to target customers based on geographic or demographic profiles. Meanwhile, the Federal Trade Commission (FTC) is evaluating mobile tracking technology as part of its ongoing mission to protect consumer privacy. It is trying to determine how far a company can go in tracking personal information before it invades customer privacy and becomes unethical.

Using Data Appropriately

Another ethical issue in marketing research is the misuse of research methods and findings. Marketing research firms may be compelled by their clients to return findings favorable to the client or to arrive at a conclusion predetermined by the client. Or marketing research firms may have employees who report false data, do not follow the directions for conducting the research, or claim credit for surveys that never were conducted.

Organizations such as the American Marketing Association, the Marketing Research Association, the International Chamber of Commerce (ICC), and ESOMAR, concerned with the reliability of research results, have established ethical standards for conducting research. The table below lists the key components of the ICC/ESOMAR standards. Such standards are important to the industry because they help gain the trust of consumers. Without that trust, individuals will be less likely to participate in marketing research. Expectations of privacy, or other ethical concerns, may vary depending on geography. By developing a global set of marketing research expectations, firms can ensure that they are adhering to appropriate marketing research standards, regardless of location.

Key Fundamentals of the ICC/ESOMAR International Code on Market, Opinion and Social Research and Data Analytics

1. Market researchers shall conform to all relevant national and international laws.
2. Market researchers shall behave ethically and shall not do anything which might damage the reputation of market research.
3. Market researchers shall take special care when carrying out research among children and young people.
4. Users' cooperation is voluntary and must be based on adequate, and not misleading, information about the general purpose and nature of the project when their agreement to participate is being obtained and all such statements shall be honored.
5. The rights of users as private individuals shall be respected by market researchers and they shall not be harmed or adversely affected as the direct result of cooperating in a market research project.
6. Market researchers shall never allow personal data they collect in a market research project to be used for any purpose other than market research.
7. Market researchers shall ensure that projects and activities are designed, carried out, reported and documented accurately, transparently and objectively.
8. Market researchers shall conform to the accepted principles of fair competition.

This Code was drafted in English and the English text is the definitive version. As the Code and ESOMAR guidelines are updated on a regular basis, please refer to www.esomar.org for the latest English text.

Clearly, it is in the best interest of companies to conduct marketing research in the most ethical manner possible. If they don't, consumers may refuse to participate in research studies, provide personal information either online or face-to-face, visit company websites, or order products online. Additionally, ethical behavior on the part of companies will make it unnecessary for the government to increase regulation of marketing research practices.

Marketing Analytics: An Insightful Look into Marketing Research Data

Differentiating between Marketing Research and Marketing Analytics

Marketing research is an iterative process that gathers data to answer a clearly defined research question; the end result provides clear outcomes and deliverables. Marketing analytics focuses on the data collected by marketing researchers and *how* those data can be used. These data typically are stored in the firm's data repository. The analytics team then examines the complete set of data gathered by the firm and may reuse them for further study, for example, to identify the effectiveness of distribution channels.

Using Data Mining to Identify Purchasing Patterns

Marketing analytics uses several methods to explore data collected through marketing research; data mining is one popular, growing method that firms use to identify and interpret purchasing behaviors. For example, several years ago Target sent a teen girl in Minneapolis coupons for maternity clothes, baby clothes, and baby furniture, which provoked her father to go to his local Target demanding to speak to the manager, to whom he angrily accused the company of trying to encourage his daughter to get pregnant. It wasn't until his daughter later informed him that she was, in fact, pregnant that the father backed down and apologized for his behavior. Although the local store manager had no idea why the Target coupons were sent to the teen, Target's marketing team knew it was because of the marketing analytics they conducted on the daughter's purchasing behavior. According to Target's statistician Andrew Pole, Target assigns every customer who visits its site a Guest ID number, which is tied to that guest's personal information, including name, credit card, or e-mail address. This information is stored in a data repository that statisticians like Pole then analyze to identify buying patterns. The teen's buying patterns aligned with that of expectant parents. These customers would purchase certain sets of products together (e.g., lotions, hand sanitizers, and vitamin supplements), and then a few months later, diapers. Target made an educated guess that this teen was also expecting and thus sent her coupons and advertisements of items an expectant parent would buy.

Statistical Modeling of Complex and Voluminous Data

Marketing research and marketing analytics work hand in hand; marketing analytics requires marketing research to provide data for further analysis and marketing analytics evaluates those data to provide deeper insight into the defined research question, in some cases, even raising additional questions for continued or additional study.

Marketing analytics have come to the forefront of marketing due to the availability of multiple large and complex data sets. Analytics assists firms in identifying relationships among these multiple data sets—relationships that are not evident from looking at marketing research studies or data alone. The data provided by online sites such as Google and Facebook, and even the firm's own website, provide a wealth of information about consumers' preferences and shopping behavior. However, these data must be organized and studied in order for them to be of any use to the firm. Data mining and data modeling are two growing fields within data science. Statistical models are used to organize and review the data. Once relationships between variables are found, then researchers must interpret the meaning of those relationships. In the Target example, if the data showed that customers who purchased items related to expectant parents also showed higher instances of purchasing blue bath mats, one may conclude that expectant parents really like blue bath mats. However, sometimes the relationships between data points are not relevant or meaningful. A more meaningful interpretation may be illustrated by the Connect Marketing Analytics exercise related to this section. In the exercise, you are asked to examine the differences between men and women in their preference for online learning. Marketing analytics can use these data to potentially validate a relationship relating to the differences in general online purchasing behavior between men and women.

Responsible Marketing

Is Marketing Research Worth the Investment?

Companies rely on marketing research to streamline their marketing efforts and effectively reach their target markets. Dove is an example of a company that continues to capitalize on its marketing research strategies and deliver advertisements that are both culturally and emotionally relevant.

Dove's "Campaign for Real Beauty" catapulted Dove from a creamy soap brand to a giant in the beauty industry. Data have shown that 68 percent of women feel the media supports unrealistic standards of beauty. Dove paid attention to the data, and in the early 2000s the company devised a campaign that focused on using real women, not models. This was a drastically different approach from other beauty companies that typically used models and air-brushing.

Over the past 20 years, Dove has continued to promote the "Campaign for Real Beauty," which has proven to be a very successful long-term strategy for the brand. In 2007 Nielsen discovered that every $1 Dove spent on advertising resulted in $4.42 in sales, which validated the investment Dove makes in marketing research.

Reflection Questions

1. What types of primary and secondary data would a company like Dove collect for its marketing campaigns?

2. How can companies use data mining to further their marketing research?

Consumer Behavior

This lesson concerns the foundations of *consumer behavior,* which is rooted in the psychology of *why, when, where,* and *how* customers purchase products. To address these questions more specifically, entrepreneurs and marketers delve into customer buying behaviors. We all make decisions for different reasons, yet we follow a very similar process to do so. Understanding the process people use to make decisions and how this process functions in different settings is one of the more interesting and useful aspects of marketing—and the focus of this lesson.

By the end of this lesson you will be able to

- Describe the stages in the consumer purchase decision process.
- Identify the influences on consumer behavior.
- Distinguish among three variations of the consumer purchase decision process: routine, limited, and extended problem solving.
- Explain how data can be used to understand the consumer decision-making process.

Marketing Analytics Implications

- Many of the models used in consumer behavior were brought over from psychology. Models such as the consumer decision-making process were developed and studied in both psychology and marketing. The data from these studies are reviewed to ensure that these models, often used to predict consumer behavior, are still accurate.
- Because consumers are influenced by a wide variety of factors, marketers must constantly study not only the data gathered from the marketing environment, but also the data gathered from the economic, demographic, political/legal, technological, competitive, and sociocultural environments in which consumers live. All these data combined help marketers determine how changes in these environments are impacting consumers.
- Marketers have developed a series of tools to help structure and assist consumers in making complex decisions.

"Why Do I Buy?": The Psychological Basis of How Consumers Make Buying Decisions

The study of consumer behavior asks *why?* Why do we buy some things and not others? Why do we buy things that we didn't intend to buy? Why do people see products differently? Why does the setting of the store change how we value a product? Asking "why?" is a powerful component of studying consumer behavior topics.

Marketers strive to understand you—the customer—to help firms better adapt their marketing mix in order to create value for you.

During the onset of the COVID-19 pandemic in 2020, consumer behavior changed dramatically. Many retailers were selling out of necessity items like toilet paper and rice, but consumers "panic shopped" other items as well. Because they couldn't get to a gym, consumers ordered weights and gym equipment for their homes, causing up to a three-month delay for some products. Pet owners rushed to order grooming supplies and nail clippers while parents of young children stocked up on bicycles, puzzles, and workbooks to make the time at home more bearable.

Consumer behaviors are likely to experience lasting changes caused by the pandemic. Consumers once reluctant to shop online or utilize curbside pickup are embracing the new shopping experience. At the same time, physical retail locations are unlikely to vanish as many consumers find happiness at the chance to touch and feel products again.

Behaviors have also changed for things like self-care, mental health, and physical wellness. Consumers are spending increased time reading and taking up hobbies—all leading to a shift in spending habits focused on experiences rather than material possessions.

Because researchers have no comparison for the pandemic, understanding how consumers will behave in the post-pandemic world is impossible. What they do know is that periods of crisis often reshape consumers' values and individual psyches and that most brands will need to update their pre-pandemic personas and communication tactics.

As many consumers feared shopping in public, online ordering and delivery services became increasingly important to both consumers and businesses as they responded to changing consumer behaviors.

The Consumer Purchase Decision Process

Consumer behavior is the way in which individuals and organizations make decisions to spend their available resources, such as time or money. Firms that understand these principles of *why, when, where,* and *how* individuals and organizations make their purchases are often able to sell more products more profitably than firms that do not.

Most consumers go through a common decision-making process that involves the following five stages:

1. Problem Recognition
2. Information Search
3. Evaluation of Alternatives
4. Purchase
5. Post-Purchase Evaluation

Stage 1: Problem Recognition

Problem Recognition Question:

What need do I have to satisfy?

Answer:

I am hungry, and I need to eat something.

Stage 2: Information Search

Information Search Question:

What products are available to meet my need?

Answer:

I had pizza at the place down the street and it was good, but let me check online for other choices.

Stage 3: Evaluation of Alternatives

Evaluation of Alternatives Question:

What product will best satisfy my need?

Answer:

I am low on money so I should go for the cheapest option.

Stage 4: Purchase

Purchase Question:

Where should I purchase from and how much should I pay?

Answer:

I will go with the place down the street because it charges only $10.50 for a large pizza.

Stage 5: Post-Purchase Evaluation

Post-Purchase Evaluation Question:

Am I satisfied with my purchase?

Answer:

It was cheap, but it was so greasy I feel sick. I have to remember not to order there again.

Note that this response exemplifies *cognitive dissonance* because the consumer is feeling regret over the purchase.

Problem Recognition

The buying process begins when consumers recognize they have a need to satisfy. This is called the **problem recognition** stage.

Imagine leaving class to find that high winds had blown one of the oldest trees on campus onto your car. You need your car to get to school, work, and social events. Because your car is destroyed, you immediately recognize that you need a new mode of transportation. In this case, due to a lack of public transportation and the distance you must travel to meet your day-to-day obligations, you need to buy a new car.

Alternately, consider the panic shopping induced by COVID-19. All of a sudden, consumers felt the need to buy large amounts of toilet paper.

Sometimes problem recognition is easy, such as when you get a flat tire. In other cases, problem recognition is more difficult, such as deciding when to begin to plan for retirement; many don't realize they need to do so until it is too late.

From a marketing perspective, it's important to keep in mind the following two issues related to problem recognition:

1. **Marketers must understand all aspects of consumers' problems, even those that are less obvious, to create products that improve or enhance consumers' lives.** For example, if marketing professionals don't know what problem, beyond the need for transportation, you want to solve by purchasing a new car, they are not likely to develop strategies that will resonate with you. Are you looking for added prestige, or do you want to spend less on monthly payments? Marketers also must recognize that consumers might be buying the same car to solve different problems.

2. **Marketers must remember that if consumers are not aware of a problem or do not recognize a need, they are unlikely to engage in any of the subsequent steps of the buying process.** For example, Whole Foods may offer free samples of a new organic coffee brand to help consumers realize that they have a problem (in this case, the problem would be that they don't own this brand of coffee that tastes better than the leading brands). For Whole Foods, helping consumers recognize how new products solve unknown needs is an important way to encourage evaluation of alternatives and potential purchase of new products.

Information Search

Once consumers recognize a problem, they begin the **information search** part of the decision process. They seek information to help them make the best possible decision about whether to purchase a product to address their problem. Consumers expend effort searching for information based on how important they consider the purchase. Consumers search for both external information and internal information.

External Information Search

When consumers seek information beyond their personal knowledge and experience to support them in their buying decision, they are engaging in an **external information search.** Marketers can help consumers fill in their knowledge gaps through advertisements and product websites. Many firms use social media to empower consumers' external information searches. For example, Ford uses Facebook, Twitter, YouTube, Flickr, and Scribd to communicate information and to deepen relationships with customers.

Consumers' friends and family serve as perhaps the most important sources of external information. Recall again the example of buying a new car and what those in your life might say about different vehicle brands or models. If your parents or friends tell you about a negative experience they had with a certain car model, their opinions probably carry more weight than those of other external sources. Reviews of a product or service by other customers also carry a great deal of weight. How often have you decided not to stream a movie, visit a restaurant, or stay at a hotel based on what others have said in reviews? The power of these personal external information sources highlights why marketers must establish good relationships with *all* customers. It's impossible to predict how one consumer's experience might influence the buying decision of another potential customer.

When buying a new car, booking a hotel room, or choosing a restaurant, consumers increasingly use sites such as Yelp and TripAdvisor to read other consumers' evaluations prior to making a purchase.

Internal Information Search

Not all purchases require consumers to search for information externally. For frequently purchased items such as shampoo or toothpaste, internal information often provides a sufficient basis for making a decision. In an **internal information search,** consumers use their past experiences with items from the same brand or product class as sources of information. For example, you can easily remember your favorite soft drink or vacation destination, which will likely influence what you drink with lunch today or where you go for spring break next year.

In our car example, your past experience with automobiles plays a significant role in your new car purchase. If you have had a great experience driving a Ford Escape or Toyota Camry, for example, you may decide to buy a newer model of that same car. Alternatively, if you have had a bad experience with a specific car, brand, or dealership, you may quickly eliminate those automobiles from contention.

Evaluation of Alternatives

Once consumers have acquired information, they can use it to evaluate different alternatives, typically with a focus on identifying the benefits associated with each product. Consumers' **evaluative criteria** consist of attributes that they consider important about a certain product. Returning to our car-purchase example, you would probably consider certain characteristics such as the price, warranty, safety features, or fuel economy more important than others. Car marketers work hard to convince you that the benefits of their car, truck, or SUV reflect the criteria that matter to you.

Marketing professionals must not only emphasize the benefits of their good or service but also use strategies to ensure potential buyers view those benefits as important. For example, Whole Foods sells its food products at premium prices, so Whole Foods focuses its marketing efforts on highlighting how its program of working with local vendors is good for the local economy and helps the global environment.

Purchase

After evaluating the alternatives, customers will most likely buy, or *purchase,* a product. At this point, they have several decisions to make. For example, once you have decided on the car you want, you have to decide where to buy it. The price, salesperson, and your experience with a specific dealership can directly affect this decision. In addition, items such as financing terms and lower interest rates at one dealership versus another can affect your decision. If you decide to lease a car rather than buy one, you would make that decision during this step.

Marketers do everything they can to help you make a decision to purchase their product, yet in the end, it comes down to many small decisions. Thus, an effective marketing strategy seeks to minimize any small decisions that could delay or derail the actual decision to purchase. Companies like Lowe's are very aware that consumers often get bogged down in details related to delivery, installation, and cost before deciding to make a purchase. For this reason, Lowe's offers a variety of services, such as design, delivery, installation, and financing. Lowe's main reason for doing this is not explicitly to earn profit from selling services, but rather to remove any roadblocks from a consumer making a purchase at its store versus a competitor like Home Depot. Because profits start with sales, providing goods and services that make it easy for consumers to make a purchase decision is one of the most important activities that marketers perform.

Post-Purchase Evaluation

Consumers' **post-purchase evaluation** occurs after the sale is complete. Consumers will consider their feelings and perceptions of the process and product, which can result in either positive or negative assessments. Post-purchase assessments are important because these evaluations about the purchase will likely impact how consumers ultimately feel about the purchase and whether or not they will purchase the product again.

Three main things are evaluated after purchase: customer satisfaction, cognitive dissonance, and loyalty. Customer satisfaction is the first evaluation.

Customer satisfaction is a state that is achieved when companies meet the needs and expectations customers have for their goods or services. If the product purchased meets expectations (for instance, it works well, it performs as described, and the buying experience was as expected), consumers will most likely be satisfied with the purchase.

In many cases, consumers experience **cognitive dissonance,** which is the mental conflict that occurs when consumers acquire new information that contradicts their beliefs or assumptions about the purchase. Cognitive dissonance, also known as *buyer's regret,* often arises when consumers begin to wonder if they made the right purchase decision. Cognitive dissonance after making a purchase can arise for numerous reasons. For example, perhaps you discover that the car you just bought doesn't get the gas mileage you'd expected or you find out someone you know bought the same car for a lower price. Marketers do various things to reduce the level of dissonance felt by consumers. A car company might offer an extended warranty, a price-match guarantee, or a no-questions-asked return policy. Reducing cognitive dissonance is important to keeping customers satisfied.

By helping consumers feel better about their purchase, marketers increase the likelihood that those consumers will remain satisfied to the point of becoming loyal to the product. **Loyalty** is an accrued satisfaction over time that results in repeat purchases. In addition, loyalty can increase consumer actions, such as positive word of mouth, reviews, and/or advocacy for the product.

Because the post-purchase period has such potential to affect long-term results, many companies work just as hard *after* the sale as they do *during* the sales process to keep the customer satisfied.

Influences on Consumer Behavior

Numerous factors affect the consumer decision-making process at every stage. These include psychological, situational and personal, sociocultural, and involvement influences. An effective marketing strategy must take these factors into account. The following sections provide an overview  of the different influences and how marketers incorporate them into their marketing activities and strategies.

Psychological Influences on Consumer Behavior

Through the course of the consumer decision-making process, consumers engage in certain psychological processes: they develop *attitudes* about a product's meaning to their lives, they *learn* about the product, and they must be *motivated* to purchase the product. Marketers need to understand each of these psychological processes as they develop strategies to reach consumers.

Attitude

An **attitude** is a consumer's overall evaluation of a product, which involves general feelings of like or dislike. Attitudes can significantly affect consumer behavior. For example, many consumers have strong positive feelings about Whole Foods because they appreciate the high-quality products it offers. Conversely, many consumers have strong negative feelings about Whole Foods, terming it "Whole Paycheck" because it charges premium prices for its products. These diverse positions represent different attitudes.

Attitudes can be both an obstacle and an advantage to a marketer. Choosing to discount or ignore consumers' attitudes toward a particular good or service when developing a marketing strategy guarantees the campaign will enjoy only limited success. In contrast, perceptive marketers leverage their understanding of attitudes to predict the behavior of consumers and develop a marketing strategy that reflects that behavior.

Learning

Learning refers to the modification of behavior that occurs over time due to experiences and other external stimuli. Consumer learning may or may not result from things marketers do; however, almost all consumer behavior is learned. Marketers can influence consumer learning, and, by doing so, impact consumer decisions and strengthen consumer relationships, but they must first understand the basic learning process.

Learning typically begins with a stimulus that encourages consumers to act to reduce a need or want, followed by a response, which attempts to satisfy that need or want. Marketers can provide cues through things like advertisements that encourage a consumer to satisfy a need or want using the firm's good or service. Reinforcement of the learning process occurs when the response reduces the need. For example, eating a pizza reduces hunger and an immediate need for food. Consistent reinforcement satisfied by a particular type of pizza (Sicilian, plain, veggie, pepperoni) can lead the consumer to make the same purchase decision repeatedly as matter of habit. Think about it, when you're hungry, do you tend to put a lot of thought into what pizza to buy or where to buy it?

Marketers can capitalize on consumer learning by designing marketing strategies that promote reinforcement. Bounty has promoted its paper towels using the slogan "the quicker picker upper" for more than 40 years. Through repetition in promotion, the company hopes to influence consumer learning by associating Bounty with the idea of cleaning up spills quickly. The strategy has worked: Bounty remains the market share leader in paper towels and continues to grow in sales volume even as other paper towel brands have faced years of declines.

Learning is important in the consumer decision-making process. When a consumer learns that a product or service satisfies a need and solves a problem, that product or service moves up to the top of the information search the next time the consumer recognizes a problem.

Motivation

A third psychological process that affects consumer behavior is motivation. **Motivation** is the inward drive people have to get what they need or want. Marketers spend billions of dollars on research to understand how they can motivate people to buy their products. One of the most well-known models for understanding consumer motivation was developed by Abraham Maslow in the mid-1900s. He theorized that humans have various types of needs, from simple needs such as water and sleep to complex needs such as love and self-esteem. Maslow's hierarchy of needs illustrates the belief that people are motivated to meet their basic needs before fulfilling higher-level needs.

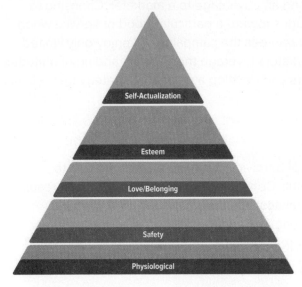

Maslow's Hierarchy of Needs has five levels organized in a pyramid shape, with levels ordered from bottom to top.

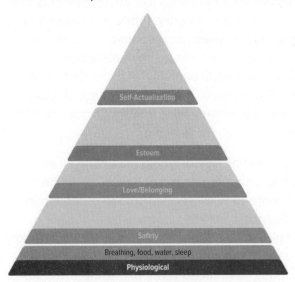

A consumer who lacks food, safety, love, and esteem would consider food his or her greatest need and, according to Maslow, would seek to fulfill that need before any of the others. Marketers of food, bottled water, and medicine are often focused on meeting the **physiological needs** of their target customers. This level represents the most basic of human needs.

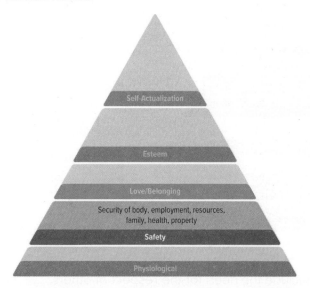

Safety can take different forms, including physical safety and economic safety. For example, the absence of economic safety—due to economic crisis or lack of job opportunities—leads consumers to want job security, savings accounts, insurance policies, and reasonable disability accommodations. Marketers match products to these types of needs because they know consumers will be motivated to buy them.

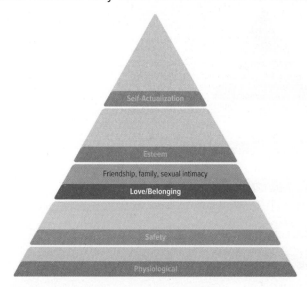

The third level of human needs, after physiological and safety needs are fulfilled, involves **love and belonging**. Love has become big business for marketers. The online dating site eHarmony was founded to serve this need by establishing its brand as the site for the serious relationship seeker.

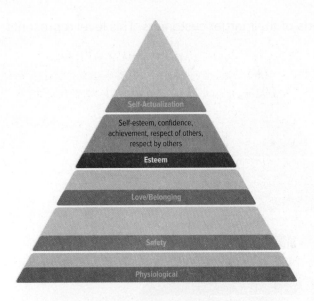

Esteem is the need all humans have to be respected by others as well as by themselves. Lower esteem needs include the need for the respect of others. Jewelry stores and luxury car makers often target their marketing at consumers with lower esteem needs or those looking to increase their status or prestige. Higher esteem needs include the need for self-respect, competence, and mastery. For example, makers of foreign language education software market their products as a way for consumers to fulfill a lifelong dream of speaking a new language.

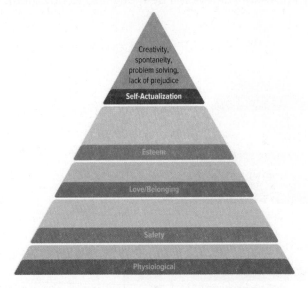

Maslow describes the top tier of the hierarchy as the aspiration to become everything that one is capable of becoming. **Self-actualization** pertains to what a consumer's full potential is and the need to realize that potential. For example, one individual may have a strong desire to become an ideal parent, another may want to become a superior athlete, and another may want to excel at painting, photography, or inventing.

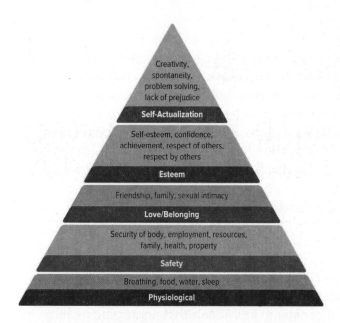

Situational and Personal Influences on Consumer Behavior

Situational and personal influences on the consumer decision-making process include factors such as time, personality, lifestyle, values, and surroundings.

Time

Consumers value their time greatly, and time considerations often affect their purchasing decisions. Companies throughout the world understand this about their customers, so they will design goods and services accordingly. For example, banks have created apps that allow banking to occur without a customer ever needing to step foot inside an actual bank (think mobile check deposits). Reducing the time a consumer spends interacting with a firm is vital in almost all consumer settings.

Time can also impact what consumers ultimately pay for a good or service. Consumers are often willing to pay more for products if the placement of those products saves them time. For example, though a consumer may realize that milk, eggs, or bread are significantly more expensive at a small, local grocery than they might be at a larger chain supermarket farther away, that consumer might be willing to pay more for these items for the convenience of shopping closer to home and checking out more quickly. By placing products in a more accessible location, marketers can often increase their profits on individual items while still providing value to their customers.

Immediacy of the need plays a role in how fast the buying process occurs. If a product is needed now, such as a replacement dishwasher, the time to elaborate on choices before making a decision will be shortened. However, if a new dishwasher is needed for the remodeling of a kitchen that is occurring in a few months, more time is available to elaborate on the decision. Often consumers are forced to skip steps in the buying process in response to an emergency or immediate need.

Personality

Personality is the set of distinctive characteristics that lead an individual to respond in a consistent way to certain situations. A consumer's personality might include traits that make them confident, personable, deferential, adaptable, or dominant. Personality strongly influences a consumer's decision to purchase products. For example, shopping at Whole Foods might demonstrate a consumer's passion for locally grown organic products, whereas shopping at Walmart might represent an individual's commitment to frugal spending. Marketing professionals strive to identify personality traits that distinguish large groups of people from each other and then design strategies that best appeal to those consumers.

Lifestyle

Lifestyle is a person's typical way of life as expressed by his or her activities, interests, and opinions. For example, Whole Foods's overall marketing message focuses on exceptional products that cater to individuals that appreciate healthy living. Lifestyle characteristics are often easier for a firm to understand and measure than personality traits because it is relatively easy to observe how consumers express themselves in social and cultural settings. Marketers can potentially reach their targeted consumers by sponsoring events related to interests or activities those consumers are passionate about, or through social media advertising.

Values

Think about how your personal value system influences how you live your life. It is likely that your value system also corresponds to your buying behavior. **Values** reflect a consumer's belief that a specific behavior is socially or personally preferable to another behavior. Personal values, which include everything from a consumer's religious beliefs to a belief in self-responsibility, can impact the decision-making process. For example, many consumers look for goods and services that embrace sustainability—that is, products intended to benefit the environment, society, and the economy. Firms take into account consumer values to design specific products, such as BPA-free water bottles or LED bulbs. Marketers then message the products as eco-friendly to appeal to customers' sense of environmental responsibility.

In this ad by clothing company Patagonia, people are told not to buy its apparel. This message was a counterintuitive attempt to appeal to consumers' concerns about environmental responsibility and sustainability.

Surroundings

Often, factors in our surroundings influence our decision-making process. Have you ever been in a grocery store that smelled bad? If you have, chances are pretty likely that you might have questioned the quality of the food products at that grocery store. Or, if you've ever had to choose between two different fitness centers, you might decide to join the one that features more modern equipment and has a more open and friendly workout space. Crowd size is another factor that can influence consumer behavior; if the crowd at the mall on the day of a huge sale is too large, it might deter you from shopping there that day. Conversely, if you notice a restaurant with just a few people dining in it during the busiest part of the day, you might be less likely to eat there.

Sociocultural Influences on Consumer Behavior

Sociocultural factors develop from a consumer's relationships with others and can significantly impact his or her buying behavior. Sociocultural influences that impact the consumer decision-making process include family members, reference groups, and opinion leaders.

Family Influences

Family members are one of the greatest influences on consumer behavior. The level of influence can vary across families and can evolve as a family ages and new members join the family through marriage or birth. The composition of families has changed greatly in recent decades to include more single parents and same-sex households, which can impact consumer decisions in different ways important to marketers. For some cultures, family is the primary influence. It can be a central theme for marketers seeking to target Hispanic market segments in both the United States and throughout the world.

Children often greatly influence a household's purchase decisions, particularly when it comes to grocery shopping and dining out. Marketers at McDonald's, Sonic, and Burger King spend a significant amount of money advertising to young consumers, giving away toys and books with their kids' meals. McDonald's has dedicated an entire website, **www.happymeal.com**, to marketing to children through games and technology. These promotions aimed at children can enhance restaurant traffic and revenues. However, marketing food to children has become a controversial topic for firms. Some suggest that marketing to children is unethical because children are impressionable and can be easily manipulated by marketing messages. Thus, marketers must practice caution when marketing to children so as not to take advantage of them.

Few situational factors have a more significant impact on consumer behavior than the family life cycle. The **family life cycle** describes the distinct family-related phases that an individual progresses through over the course of his or her life time. There are six stages that an individual might go through as part of his or her family life cycle:

1. Unmarried

2. Married with no kids

3. Married with small children or tweens

4. Married with teens

5. Married without dependent children

6. Unmarried survivor

Each stage impacts consumer behavior, and therefore a firm's marketing strategy to that consumer. Companies like State Farm Insurance actively promote their ability to service consumers throughout the entire family life cycle. They promote products and services that help customers insure their first car, secure family life insurance, and invest for retirement. In addition to family, consumers typically belong to or come into contact with various other social groups—schools, workplaces, churches, and volunteer groups—that can influence their purchase decisions.

1. Single

Single

This stage is characterized by **individual spending** (such as fitness, fashion, dating, travel, hobbies, etc.).

2. Married

Married

This stage is characterized by personal **couple's** spending (travel, self-improvement, work clothing, fine dining, education, etc.).

3. Young Family

Young Family

This stage is characterized by **family spending** on infants and toddlers (such as diapers, baby food, clothing, safety equipment, daycare, etc.) and tweens (sports equipment, summer camps, fast food, clothing, etc.).

4. Family with Teens

Family with Teens

This stage is characterized by spending on **teens** (such as college funds, cars, fashion, travel, technology, etc.).

5. Empty Nest

Empty Nest

This stage is characterized by **future spending** (such as travel, retirement planning, first homes, cars, etc.).

6. Survivor

Survivor

This stage is characterized by **freedom spending** (such as travel, hobbies, grandchildren) and **health** (doctor fees, insurance, medication, etc.).

Reference groups can provide consumers with a new perspective on how to live their lives. When you accept your first job, how will you know what to wear on your first day at work? You might recall what you saw people around the office wearing when you interviewed. In such a case, your coworkers serve as your reference group. A **reference group** is made up of the people to whom a consumer compares himself or herself. Marketers should understand that the more public the purchase decision, the more impact reference groups are likely to have. For example, reference groups tend to significantly influence a consumer's clothing purchases.

Firms typically focus on three consumer reference groups when developing a marketing strategy: **membership reference groups, aspirational reference groups,** and **dissociative reference groups.**

Membership Reference Group

A membership reference group is the group to which a consumer actually belongs. Membership groups could include school clubs, fraternities and sororities, and the workplace.

Marketers who understand the influence other members of these groups have on consumers can target products that would be ideal for the group members. For example, a local bank might market itself as the official bank of your university and offer a debit card featuring the school's logo.

Aspirational Reference Group

An aspirational reference group refers to the individuals a consumer would like to emulate. For example, professional athletes represent an ideal for many people. Serena Williams, who was ranked world No. 1 in singles by the Women's Tennis Association from 2002 to 2017, is one such iconic sports figure. She is featured in an inspirational Gatorade ad speaking to her newborn daughter, and in so doing, delivering a greater message to all young girls interested in sports, or aspiring to be an athlete: "Whether your bond is by blood or by ball, whether she shares the color of your skin or the color of your jersey, you'll find sisters in sweat."*

Dissociative Reference Group

Dissociative reference groups include people that an individual would *not* like to emulate. Teenagers and young adults tend to actively dissociate themselves from groups they view as "uncool" or in which their parents might express interest. But dissociative reference groups can play a role in marketing to all consumers. Marketers for mouthwash and certain types of chewing gum encourage consumers to use their products as a way to avoid being associated with those who have bad breath.

Opinion Leaders

Individuals who exert an unequal amount of influence on the decisions of others because they are considered knowledgeable about particular products are called **opinion leaders.** Opinion leaders range from Michael Jordan endorsing a pair of Nike shoes to Rachael Ray promoting a specific type of cooking utensil.

Opinion leaders are not just celebrities; social media allow small numbers of consumers to influence the consumer decisions of a much larger group. More marketers are trying to tap into the power of these social media opinion leaders through rewards and other benefits. For example, credit card companies offer special rewards to customers who have the potential to influence others; airlines give these same types of opinion leaders free flights in an effort to encourage them to use their influence on behalf of the company's products.

Involvement Influences on Consumer Behavior

How consumers make choices is influenced by their level of involvement in the decision process. **Involvement** is the personal, financial, and social significance of the decision being made. The study of high and low involvement focuses on how consumers choose which alternative to purchase and is important for firms to understand as they develop strategies to sell their products.

Characteristics of Purchasing Decisions

	Low-Involvement Purchase	High-Involvement Purchase
	Usually inexpensive products that pose a low risk to the consumer if the purchase ends up being a mistake.	More significant purchases that carry a greater risk to the consumer if the purchase ends up being a mistake.
Price	inexpensive	expensive
Preparation	requires little forethought	requires research
Frequency	frequently purchased	rarely purchased
Risk	limited risk	high risk

Low-Involvement Buying Decisions

Most likely, you have made an impulse purchase sometime in the past month. Impulse buying is purchasing a product with no planning or forethought. Buying gum in a grocery store checkout line or a new cap that you notice as you walk through a mall are examples of impulse buying. Impulse purchases usually occur with low-involvement products. **Low-involvement products** are inexpensive products that can be purchased without much forethought and with some frequency.

Consumers often do not recognize their desire for a low-involvement product until they are in the store, which influences the strategic decisions for marketing these items. In-store promotion, for example, is a very useful tool for marketing low-involvement products. Unique packaging or special displays help to capture the consumer's attention and quickly explain the product's purpose and benefits. Marketing strategies for low-involvement products include Kellogg's signage at Walmart stores explaining the relatively low cost of eating breakfast at home versus at a restaurant, or a promotional sign by Sara Lee in a grocery store aisle that focuses on a promotion for its "soft and smooth bread." Tactics like low-tech cardboard displays found at the end of aisles can potentially drive more impulse purchases than temporary price reductions.

High-Involvement Buying Decisions

High-involvement products include more significant purchases that carry a greater risk to consumers if they fail. Two common examples of high-involvement purchases are a car and a house. Companies that market high-involvement products must provide potential consumers with extensive and helpful information as they go through the decision-making process. An informative advertisement can outline to the consumer the major benefits of a specific product purchase. Residential brokerage firm Coldwell Banker provides a wealth of information about the homes it is attempting to sell to help potential buyers understand more about the house itself as well as about local schools, financing options, and moving services.

Variations of the Consumer Purchase Decision Process

Marketers must be aware that there are three variations of the consumer buying process. All start with recognition of a need, but follow a slightly different path based on if the consumer purchase is the result of a *routine, limited,* or an *extended* problem-solving decision. For regularly purchased products, the consumer will not go through all the steps of the decision process. These products are typically low-involvement purchases. On the other extreme, for first-time purchased products, the consumer will engage in extended problem solving and most likely go through all the steps. These purchases require more involvement from the customer. The differences between these variations are highlighted in the table that follows.

Variations of the Consumer Buying Process

	Routine Problem Solving	Limited Problem Solving	Extended Problem Solving
Example	**Toothpaste**	**New Cell Phone**	**New Car**
Frequency of Purchase	Regularly	Occasionally	Infrequently or never
Experience in Purchasing	Experienced	Some experience	None or little experience
Level of Involvement	Usually low level	Mid-level	High level
Coverage of Decision-Making Stages:			
Problem Recognition	Yes	Yes	Yes
Information Search	No	Limited amount	Yes
Evaluation of Alternatives	No	Limited amount	Yes
Purchase	Yes	Yes	Yes
Post-Purchase Evaluation	Limited amount	Yes	Yes

Marketing Analytics: Using Data to Understand Consumer Behavior

Using Research to Develop Predictive Models of Consumer Behavior

The models used in consumer behavior were developed primarily in psychology and were based on theories of how people think and behave. In marketing we are primarily interested in how consumers think and behave in a *marketplace*. Firms use data to convert these theories into workable models that could be used to predict how consumers will respond to market offerings. Often to evaluate these models fully, ongoing data collection must be conducted. For example, consumer satisfaction measures are often created from survey data. Consider the surveys you might find at the end of a receipt from a store or restaurant. The data from these surveys inform the consumer's post-purchase evaluation stage and give firms the information they need to ensure they are fully satisfying a consumer's needs.

Understanding Influences on Consumers

What motivates a consumer to choose one product over another is a complex process that involves both physical and psychological states. These states are influenced by the broader environments in which consumers live. To truly understand these influences, marketers must collect and analyze data from these different environments, including the economic, demographic, political/legal, technological, competitive, and sociocultural environments.

Improvements in technology have changed how consumers interact with the market—for example, data show that more and more consumers are shopping online versus at brick-and-mortar stores. Economic data trackers, such as the Consumer Confidence Index, allow marketers to better understand how consumers are feeling about making big purchases. Marketers must stay on top of changes in the political/legal environment as well, and consider the impact these changes have not only on consumer behavior, but also on consumer access. For example, in some states consumers are able to order wine from online merchants, whereas in other states this is against the law. Regulations can be written that prohibit or limit consumer's access to such products, so marketers must take this into consideration when making marketing decisions. The physical environment, such as global climate change, is also influencing consumer behavior. Data reveal a growing interest in the purchase of more sustainable products—from automobiles to food to clothing and even travel. Using data that track the changes in these environments allows marketers to make predictions about what will happen next for consumers. It also adds a new dimension to their strategic planning.

Models of Consumer Decision Making

Not only can marketers use consumer decision-making models to understand consumers, but consumers can also use these models to help *them* make marketplace decisions. For routine or known problem solving, most consumers use an established method of decision making; they buy whatever they bought before (based on previous experience) to solve their market problem. In such cases, they often do not evaluate other alternatives. This form of decision making is called **noncompensatory decision making.**

For novel problems that involve some level of risk, consumers will consider multiple decision-making criteria and a more complex decision-making process. This form of decision making is called **compensatory decision making.** In this model, consumers assign **weights,** measures of importance or preference, to the various criteria they've predetermined. Assigning weights—for instance, on a scale of 1 to 5, with 5 being most important and 1 being least important—is a way for consumers to evaluate how different product options compare overall to each other.

The weights that consumers assign to the same set of criteria will vary with each consumer. Let's consider the criteria involved in purchasing a car: price, comfort, fuel economy, and style. Consumer A may rate style as more important than price, whereas Consumer B may rate price as the most important criterion of all. The sum (or score) of these weights, known as the **weighted preference,** will be used to make a purchase decision. After completing this exercise, some consumers may find that the product they thought they originally wanted is now, in fact, no longer the same product they end up buying. For example, a couple planning a wedding may have decided they wanted to marry in a specific location because it holds special meaning for them. They have three other venues in mind as well. They evaluate all their options using the compensatory decision-making model, considering location, cost, ease of access for guests, and catering options. After assigning weights to each of these criteria and calculating the weighted scores, the couple realizes that their original choice of venue is not the best option for them after all.

Responsible Marketing

Patagonia and Consumer Decision Making: Sustainability and Beyond

During the consumer purchase decision process, certain psychological, situational, personal, sociocultural, and involvement influences can affect a consumer's decision to purchase. What mood is the customer in? What values does the customer have that affect shopping behavior? What is the state of the economy? With so many options available to consumers and so many factors affecting purchasing decisions, it can be difficult for companies to convert sales. One company that has grown its customer base and continues to attract customers with shared values is Patagonia.

The text references Patagonia and its unusual campaign strategy in 2011, which told people to not buy its apparel. This campaign was designed to bring attention to the company's position on sustainability (ironically, Patagonia sales increased by 30 percent that year). In subsequent years, Patagonia has amplified its mission and promotion of sustainability. The company resells gently worn items on its website and encourages customers to turn in used products for credit. This is in keeping with its customers' values; a 2017 survey indicated that 69 percent of Patagonia customers said they were concerned where the products come from.

While Patagonia has increased awareness of sustainability, COVID-19 has greatly impacted consumer behavior. Euromonitor reported that "73% of professionals believe sustainability efforts are critical to success"; however, with many customers forced to order merchandise online during the pandemic, 51 percent reported an increased use of plastics. How Patagonia will deal with these and other challenges to sustainability in the future remains to be seen.

Reflection Questions

1. Consider the five steps in the consumer purchase decision process. What are considerations a Patagonia customer might make in each step?

2. What are some potential psychological, situational, personal, sociocultural, or involvement influences on Patagonia customers in their decision-making process?

3. How has COVID-19 impacted consumer behavior and how may it affect Patagonia's message going forward?

Business-to-Business Marketing

This lesson will explain how businesses that sell to other businesses market their products and services. Marketing to business customers involves some similar and many different types of activities than those used in B2C marketing. For example, as a consumer, you may be familiar with Champion products. However, have you considered the importance of the retail relationships that Champion has to develop to have a successful brand?

By the end of this lesson you will be able to

- Describe the buyer–seller relationship.
- Describe the different types of buying situations.
- Recognize the major forms of B2B e-marketing.
- Outline the steps in the B2B buying process.
- Define buying centers, how they influence organizational purchasing, and the roles of their members.

Champion: Becoming "Cool" Again

Have you ever heard of Champion? Originally called the Knickerbocker Knitting Company, Champion was founded as a family business in 1919. It started out selling athletic uniforms for the University of Michigan and gained traction with its durability and comfort. Growing in style and notoriety, Champion obtained partnerships with the National Collegiate Athletic Association (NCAA) in the 1960s, the National Football League (NFL) in the 1970s, and the National Basketball Association (NBA) in the 1990s.

After years of partnerships with collegiate and professional sports, demand for the brand waned. Consumer demand shifted from what turned into the "frugal dad" look to new brands such as Lululemon and Under Armour.

So how did Champion become trendy again? For starters, in 2006 the Sara Lee Corporation spun off HanesBrands to contain all of its clothing brands, including Champion, when it no longer had time for "niche businesses" and became fully invested in the food industry. Champion started partnering with trendy brands such as Supreme, UNDEFEATED, and Vetements with the mindset that if it wanted to become "cool" again, it had to partner with trending brands.

As Champion began making its comeback, vintage clothing and athleisure were also becoming popular. Thus, products designed around these trends were created for maximum hype and designer brand partnerships. Currently focused more on pop culture than sports, Champion clothing can be found in trendy clothing stores such as Zumiez, Urban Outfitters, PacSun, and more.

So who is HanesBrands, and how does it get Champion products into the hands of consumers? HanesBrandsb2b.com is the business-to-business website that business partners can order from directly. This wholesale website puts business customers in control of their order. Hanes makes it easy for businesses big or small to order with no minimum order quantity, receive fast and affordable shipping, chooses from drop shipping options, and have access to bestselling brands.

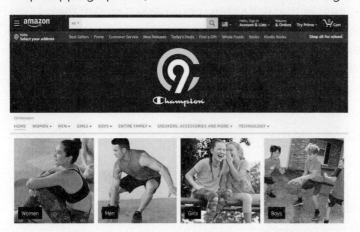

In March 2020, HanesBrands teamed up with Amazon offering the C9 Champion performance athletic wear as part of a multiyear partnership. More than 100 styles from C9 Champion will be available via Amazon's online store, providing consumers with a new channel to access the brand. HanesBrands signed a multiyear agreement for the C9 Champion Athleticwear brand to be sold exclusively through Amazon.

Factors Affecting Business-to-Business Marketing

As noted in the introduction to this lesson, **business-to-business (B2B) marketing** consists of marketing to organizations that acquire goods and services in the production of other goods and services that are then sold or supplied to another business or consumer groups.

B2B marketers face many of the same challenges B2C marketers do, as both work to connect or build relationships with their customers. However, unlike in B2C marketing, where marketing is targeted at individual consumers, in B2B marketing, marketing is targeted at **professional purchasers or buyers**—employees of companies who make purchase decisions in the best interest of their organizations. Because of this, B2B marketing is based on long-term relationships that are often maintained among multiple people from the selling firm and a group of people from the buying center of a purchasing company. In addition, B2B buyers and sellers are motivated by derived demand for products rather than individual decision making based on personal needs and wants. Let's look at each of these factors in a bit more detail.

Professional Purchasing

Businesses typically buy things through professional purchasing managers who are experienced in the policies and procedures necessary to make a large deal. For example, a professional buyer at Old Navy will be responsible for purchasing the clothing styles and accessories that will eventually be featured in Old Navy's stores. Because businesses strategically plan well into the future, the B2B purchasing process is often far longer than the consumer decision-making process. In addition, the B2B buying process requires standardized procedures, such as a request for proposal (RFP) and contract negotiations, which are not typically found in consumer buying. Finally, the strategic nature of business purchasing removes much of the emotion and personal interest from purchase decisions. For example, Sysco is a large wholesale distributor of food and beverage products. A buyer–seller relationship at Sysco may involve discussion of the sale/purchase of 200 cases of tomato soup for a large catering event. It should be easy to see how the buyer of a large quantity of soup will be more strategic and less emotionally invested in the purchase than he or she would be in a personal decision to buy a new iPhone or concert tickets.

Importance of the Buyer–Seller Relationship

In addition to buyers being professionals, business marketers typically deal with far fewer buyers than consumer marketers. Usually, each of these customers is larger and more essential to the firm's success than a consumer because there are fewer business buyers. For example, the potential demand for pizza is almost unlimited in the United States. However, the demand for large-scale pizza ovens is confined to medium and large pizza establishments, many of which belong to Pizza Hut, Little Caesars, and other major national chains.

Because there are fewer buyers, B2B marketers feel even more pressure to make sure they offer high-quality products to and establish good business relationships with their customers. A B2B **buyer–seller relationship** is a connection between a firm and/or its employees intended to result in mutually beneficial outcomes. B2B buyer–seller relationships are critical because a bad relationship with an individual pizza consumer might cost the local Pizza Hut $20 per week, whereas a bad B2B relationship with Pizza Hut might cost a pizza oven maker its entire annual profit or even its future.

Derived Demand

The need for business goods comes from demand for consumer goods. **Derived demand** occurs when demand for one product occurs because of demand for a related product. Even though Sysco, for example, does not sell directly to consumers, the success of its business depends on the buying patterns of individual consumers. Derived demand also provides an important reason to develop mutually beneficial relationships with B2B partners.

Imagine a scenario in which Sysco has a contract as the only provider of food to campus dining at your school. Because Sysco is the only supplier, it has the option to charge far more than is necessary for the goods it sells, thereby maximizing profit. However, if Sysco chooses to charge your university higher prices, the university, in turn, will have to pass on those cost increases to students like you. Once the price of campus dining gets too high, you and other students will simply find other places to eat. In turn, overall student demand for cafeteria meals will decrease, thereby decreasing the university's need for Sysco's products.

This scenario illustrates why marketers must take a strategic view of business relationships and understand all of the potential impacts their actions can have on derived demand.

B2B marketing professionals focus on and maintain relationships with buyers who work in different business markets. In addition, within each market buyer–seller relationships can vary. Let's discuss different business markets and the continuum of business relationships.

Reseller Markets

Resellers include retailers and wholesalers that buy finished goods and resell them for a profit. A retailer, such as a clothing or grocery store, sells mainly to end-user consumers like you. A **wholesaler** is a firm that sells goods to anyone other than an end-user consumer. Sysco is a perfect example of a wholesaler. Wholesalers frequently purchase a large quantity of a good (e.g., 200,000 pounds of hamburger meat) at a low cost and then sell off smaller quantities of the good (e.g., 20-pound cases of hamburger meat sold to restaurants) at a higher per-unit price. Thus, wholesalers are often called intermediaries because they don't produce products, they just process them.

Government Markets

Government markets include thousands of federal, state, and local entities that purchase everything from heavy equipment to paperclips. The U.S. government is one of the world's largest customers, spending billions of dollars a year. Marketing goods and services to the U.S. government requires adherence to certain policies, procedures, and documentation obligations. Because the public holds the government accountable for its purchases, complex buying procedures are often used to ensure that purchases meet the necessary requirements. Firms must be detail oriented and complete extensive documentation to succeed at marketing to federal agencies and departments. For example, Mississippi-based Gulf Coast Produce spent considerable time and resources winning a government contract to provide millions of dollars in fruits and vegetables to the military. This complicated and often slow process has made some marketers, especially in small businesses, reluctant to bid on government business. Of the 20 million small businesses in the United States, only about 500,000 of them have completed the documentation necessary to be eligible to sell to the government. However, government markets can be highly lucrative for smart marketers and organizations.

Institutional Markets

Institutional markets represent a wide variety of organizations, including hospitals, schools, churches, and nonprofit organizations. Institutional markets can vary widely in their buying practices. For example, a megachurch with thousands of members and a multimillion-dollar budget will likely have a buying manager for firms to work with, whereas marketing to a new church with a small congregation might simply require speaking with the pastor. These diverse buying situations pose unique challenges for institutional marketers. They must develop flexible, customized solutions that meet the needs of differently sized organizations. Educating institutional customers about how specific goods and services can make their organizations more efficient or effective is a firm's best tool for selling products in this type of market. For example, marketers for a medical technology firm could show a hospital how their customized technology solutions can reduce costs for the hospital while improving patient care.

Reseller Market

Nordstrom is an example of the reseller market. Nordstrom processes products as they come in through a distribution center, and resells them by such brands as Madewell, Patagonia, and Clinique to the end consumer.

Government Market

The government is a huge user of food products. The Armed Forces, federal cafeterias, and even the White House all purchase food on behalf of the government. Sellers to government commissaries (foodservices) must comply with complex and specific regulations to do business with the government.

Institutional Market

School lunch is just one example of an institutional market for food products. Though many schools and institutions are government run, they are a category of their own. Schools, hospitals, and nursing homes are regulated by different organizations and standards that require sellers to comply with specific guidelines and policies.

Continuum of Buyer–Seller Relationships

B2B relationships are often described as being somewhere on a continuum, which means they vary but fall within a range of two different types:

- **Transactional relationships:** On one end of the continuum, relationships are transactional, which means they are based on low prices, commodity products, and are not usually stable in the long term. The pure **transactional relationship** exists where both parties protect their own interests and where partners do things for each other purely as an exchange.

  ▶ Many organizations, for example, purchase office supplies like printer paper, staples, paper clips, and pens. While the organization may have a relationship with the provider they work with, many of these purchases are just transactional. Organizations mostly look for the best price for these types of products.

- **Collaborative relationships:** At the other end of the continuum, relationships are collaborative, which means that both parties in the relationship are highly engaged and dependent on working with each other for a long time. The **collaborative relationship** occurs when both parties share resources (e.g., financial risk, knowledge, and employees) in an effort to attain a common goal that provides beneficial outcomes to both parties.

  ▶ It is important for Champion to work closely with its retail partners. The success of the retailer and the success of Champion are often tied together. Champion is dependent on its retail partners to advance and build the Champion brand. As retailers are often the final interface with consumers and advocates for the brands they sell, Champion must provide its retail partners with information and materials to educate the consumers they serve.

Most B2B relationships fall somewhere between transactional and collaborative. In many cases, one firm may act as a transactional partner for some products and a collaborative partner for others. Thus, selling firms must evaluate the relationships they maintain with their buyers and invest their resources accordingly.

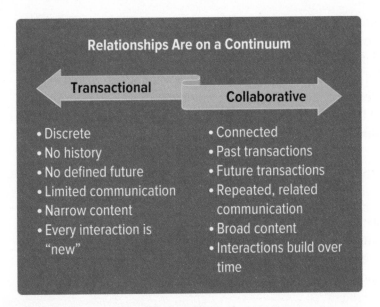

Relationships Are on a Continuum

Transactional ← → Collaborative

Transactional:
- Discrete
- No history
- No defined future
- Limited communication
- Narrow content
- Every interaction is "new"

Collaborative:
- Connected
- Past transactions
- Future transactions
- Repeated, related communication
- Broad content
- Interactions build over time

B2B Relationship Continuum. This continuum shows how B2B relationships vary but fall within a range of two different types: transactional and collaborative.

Buying Situations

Although the types of B2B customers can vary, the buying situations for each are often quite similar. Marketers can classify B2B buying situations into three general categories.

A **new buy** involves a business customer purchasing a product for the first time. For example, let's say that Dell is looking to market its personal computers to a college that has not previously bought from the company. Because the college has little or no experience with Dell, its decision process will likely be extensive, with the college requiring a significant amount of information and negotiation. From a marketing standpoint, Dell's reputation for meeting specifications and providing high-quality service to its current business and college customers could prove to be a critical factor in selling to the college for the first time.

A **straight rebuy** occurs when a business customer signals its satisfaction by agreeing to purchase the same product at the same price. B2B marketers prefer the straight rebuy outcome to any other because straight rebuys normally do not require any additional design modifications or contract negotiations. Another major advantage of a straight rebuy is that the customer typically does not look for competing bids from other companies. To revisit the Dell example, in a straight rebuy scenario, Dell should work hard to produce high-quality computers at a competitive price and offer great service that makes the college feel good about its purchase decision. Marketers who make it as easy as possible for customers to do business with their firm increase the likelihood that the customers will perceive value and develop loyalty.

A **modified rebuy** occurs when the customers' needs change or they are not completely satisfied with the product they purchased. In our Dell example, the college might want Dell to modify its computers to add additional features, lower its prices, or reduce delivery times to get new products to the school. Modified rebuys provide marketers with both positive and negative feedback. By buying from Dell again, the college signals that it is pleased with at least certain parts of its purchase experience. However, modified rebuys can also be negative if the college asks Dell to reduce its price or modify design characteristics to a point where the agreement no longer earns Dell a profit. If Dell agrees to terms that cause the company to lose money, its long-term health as an organization could be in jeopardy. Thus, the ability to walk away from potential modified rebuys provides marketers with an important tool that should be used to negotiate a positive solution that works for both the buyer and seller.

Regardless of what type of business customer B2B marketers are selling to or what type of buying situation they're in, B2B marketing should seek to create, communicate, and deliver value to customers in a way that is ultimately profitable, just as marketers would with individual consumers.

New Buy

For Sysco, *new buys* will come from new customers or by expansion of what an existing customer purchases. Acquiring new buys takes much of a salesperson's time and effort. However, once a new buy is established, it becomes a *rebuy,* which takes less effort and is the basis of a seller's recurring business portfolio. Thus, much time is spent developing new buys with new or existing customers.

Straight Rebuy

For *straight rebuy* situations, like weekly ordering at a restaurant to replenish inventories, Sysco foodservice provides an online and mobile application. Because straight rebuys require the least amount of effort for both the buyer and seller, e-commerce is an effective way to accomplish them without extra hassle.

Modified Rebuy

A *modified rebuy* for a Sysco salesperson may require that specialists in menu planning or meal preparation be consulted to resolve any issues with the original product order. Sysco invests heavily in test kitchens and services to help restaurant chefs find the right solution and taste portfolio for their customers.

The Forms of B2B E-Marketing

B2B marketers use several digital marketing tools to reach their customers, many of which are very similar to consumer marketing, and many that are not. Major B2B e-marketing tools include websites, mobile applications, e-commerce platforms, social media, and other forms of communication.

Websites

B2B website marketing is similar to consumer website marketing. B2B companies are expected to have a quality website that contains useful information concerning the company's products, services, and contacts. B2B marketing websites do pay special attention to a few additional things, often not included in consumer websites. For example, B2B websites may contain highly technical information and specifications concerning the selling firm's products. Inclusion of deeper levels of information and specifics of a product is useful to professional buyers who require more specific information than is typically available to consumers. Champion is a good example of a company that provides its retailers with a great deal of information about its prodcuts. This helps the retailers sell Champion products better.

Mobile Applications

In addition to websites, B2B sellers offer mobile applications that help their customers have real-time access to their accounts and orders on a mobile platform. Integration of a supplier's mobile technology into a buyer's supply chain flow is a good way to keep everyone in the process (e.g., purchasing, logistics, accounting, and manufacturing) informed. For example, Sysco's mobile application allows customers to access inventory levels in real time, track delivery status, review accounting statements, and even message with delivery drivers. Having a value-adding, time-saving application in a B2B setting is becoming a way that selling firms can create competitive advantage over those vying for a customer's business.

E-Commerce Platforms

E-commerce allows buyers and sellers to conduct exchange through a digital channel. For B2B firms, e-commerce is most closely associated with standard rebuy situations and transactional relationships. B2B sellers have traditionally facilitated exchange through a salesperson who meets with a buyer on a regular basis in an effort to collect orders and communicate them to the selling firm. In many cases, e-commerce eliminates this function of salespeople. However, most salespeople look at this change as positive because it allows them to focus on more important customer needs, such as customized solutions and new buys, or modified rebuys, while less complex orders process electronically. One important consideration for B2B e-commerce is security. This is similar to consumer e-commerce, but B2B e-commerce security is concerned with cyberattack and competitor security. To reduce security risks, most B2B e-commerce sites are accessed through permission of the selling firm that is granted only to customers.

Social Media

B2B firms use social media to keep their customers informed and to learn about their customers. In many B2B settings, social media have become an effective way for customers to interact with each other as well as interact with selling firms. The use of forums, blogs, and groups allows engagement to occur around a firm or product. This type of engagement, especially B2B customer-to-customer engagement, is helping firms get more immediate feedback on their products and customers to learn new ways to use a product or to solve a problem. For example, Sysco Foodservice has a LinkedIn page that has more than 90,000 followers, which allows Sysco to disseminate new information about the company quickly.

Other Forms of Communication

New forms of e-marketing are helping B2B marketers be more in touch with their customers than ever. **Enablement tools** are applications that streamline buyer–seller engagement in an effort to improve the customer's experience. Customer relationship management (CRM) systems are the basis of enablement tools. CRM allows sellers to manage their relationships with customers more effectively by closely tracking the activities and needs of customers. Enablement tools use CRM information to automate communications and follow up on customer needs. Today, tasks that once were time-consuming sales and marketing burdens are handled through enablement tools. From proactively predicting customer orders to delivery reminders, enablement tools are changing how and when B2B firms communicate with their customers in an effort to keep customers highly engaged.

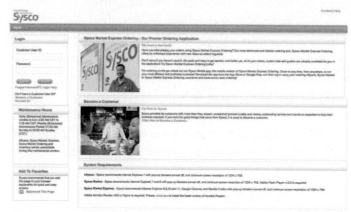

The B2B Buying Process

The B2B buying process involves research and strategic planning. As information and access to specific seller details have become more available to B2B buyers, the timing of when a seller becomes involved has changed. Sellers used to be involved in the strategic process early, as a conduit to needed information (specifications, prices, availability, etc.). Today, sellers often become involved in the process much later, with some firms estimating as late as 70 percent of the way through the process. This change in where and whom the information comes from is reflected in the process that professional buyers follow to make B2B purchase decisions, which we show in the figure below and discuss next.

STEP 1	STEP 2	STEP 3	STEP 4	STEP 5	STEP 6	STEP 7
A need is recognized	The need is described and quantified	Potential suppliers are researched	Requests for proposal are sought	Proposals are evaluated and suppliers are selected	The order is placed	A post-purchase review is conducted

Step 1: A Need Is Recognized

At some level of the organization, a need is recognized. Need recognition can come from any functional area, for example, new product development, operations, and/or information technology. New product development typically generates needs as a new product is developed, such as a new plastic formulation, a new type of product label, or an innovative sealing process. Operations often realize needs in the area of manufacturing machinery, plant safety, or energy efficiency. Information technology often identifies needs such as new software, hardware upgrades, or security enhancements. All of these needs eventually are communicated to the department in charge of purchasing, where the procurement process starts.

Step 2: The Need Is Described and Quantified

The buying center, or group of people brought together to help make the buying decision, considers the need that has been communicated and develops parameters of what needs to be purchased. In some cases, this is easy (rebuy) because a good deal is known about what is needed. In other cases, this is difficult (new buy) because what is needed is totally new and different from what is normally purchased by the firm. Specifications of the needed product are determined, such as features, amount required, where it is needed, and method of delivery. Technical or complex products require users of the new product to define the product's technical specifications. When the details are well defined, the type of product will be clear—does this product already exist or does it need to be custom made?

Step 3: Potential Suppliers Are Researched

At this stage, purchasers seek information about the product they need and the vendors that can supply the product. Online research allows most of this stage to be completed without contacting potential suppliers. Purchasers may also investigate industry forum websites and blogs of industry experts. Professional buyers often play a key role when it comes to deciding which vendors are the most qualified. Some strategic questions are also addressed in this stage, such as: Is the supplier reliable and financially stable? Where is the supplier located and will logistics be an issue? Does the supplier have enough inventory or capacity to deliver what is needed? Typically, at the end of this stage only a few qualified vendors are identified as probable suppliers.

Step 4: Requests for Proposal Are Sought

The next step is to get supplier quotes for supplying the needed product. It is important to note that to this point in the process, many purchasers will not have contacted a seller. In the past, suppliers would have been consulted much earlier in the process, but often they are now first contacted when most specifications are already decided. A **request for proposal (RFP)** is a formal listing of specifications for a needed product that is sent to a supplier firm asked to bid on supplying the product. The RFP clearly states what is needed and asks the supplier to submit a proposal for the business that includes specifics on factors such as quality, price, financing, delivery, and after-sales service. Most RFPs also include a strict deadline that indicates when competing proposals will be considered. Because in many cases this is the first time the supplier hears about the need, good sales and marketing professionals go deeper than just providing what is asked for in the RFP. Rather, selling firms use the RFP to investigate the buyer's problems and needs and how to adapt their offers to solve those problems and needs.

Step 5: Proposals Are Evaluated and Suppliers Are Selected

During this stage, the supplier responses to RFPs are reviewed and ultimately one supplier is chosen. Companies evaluate supplier responses differently, but all use some formula to evaluate what value will be realized from each supplier's offer. Many things can indicate value differently, such as high levels of service or quality that one firm offers but others do not, or innovative solutions. In the end, the lowest-priced bid may not be the one chosen, as many variables are considered in determining the best value.

Step 6: The Order Is Placed

Once a supplier has been selected, the agreement is finalized. Finalizing an agreement can be simple or complex depending on the buying situation. For example, a restaurant that decides to buy taco shells from Sysco likely just needs to complete some simple paperwork. However, a government agency that decides to purchase a missile system from a defense contractor likely involves a team of lawyers to finalize the details of the sale.

Step 7: A Post-Purchase Review Is Conducted

Once a purchase is complete, buyers conduct a review of how well the order was fulfilled. The review can be simple—did the order arrive on time and correctly?—or it can be a complex and formal process. Regardless, "getting it right" is important for suppliers that desire to continue supplying the needs of a customer.

The Influence of Buying Centers and the Roles of Their Members

Many buying decisions are complex and require decisions by multiple members of a buying center. A B2B **buying center** is a group of people responsible for strategically obtaining products needed by the firm. The buying center is everyone involved in any way in the B2B buying process. This includes the purchasing manager who places the order, the manager or executives who approve the purchase, and anyone who recognizes a purchase need. Buying center roles are informal and are not built into the company's corporate hierarchy. It is important that sellers be able to determine all the decision makers in the buying process. In an engineering firm, for example, the engineers may be the ones making buying decisions; in a goods production firm it may be product managers, and in a small company almost everyone may be involved.

Because many people are involved in buying centers, it is important to understand the roles they may play in purchase decisions. Several different general buying center roles exist. Let's discuss the role of each.

Users

Users are the individuals who will use the product once it is acquired. Users often make specifications about what the product needs and start the buying process.

Gatekeepers

Gatekeepers control the flow of information into the company that all other users review in making a purchasing decision. A personal assistant can be a gatekeeper by deciding who gets appointments with buyers and deciders.

Deciders

A **decider** is the person who chooses the good or service that the company is going to buy. The decider may not actually make the purchase but decides what will be purchased. This role can be filled by many people ranging from the CEO of the company in a large purchase decision, to the head engineer on a project, to an individual office manager buying office supplies.

Influencers

Influencers affect the buying decision by giving opinions or setting buying specifications. IT personnel and those with special knowledge on projects are often important influencers.

Buyers

Buyers are those who submit the purchase to the salesperson. This role is often more formal, such as a purchasing manager. Many firms also have specialized buyers, trained to help negotiate for the best prices of materials for the firm.

The buying center can be different across firms, and even across decisions within firms. Thus, not every role is present in all purchasing decisions. However, understanding who is performing which role, and which roles are a factor in decision making, is critical to firms that wish to effectively win a sale of their product.

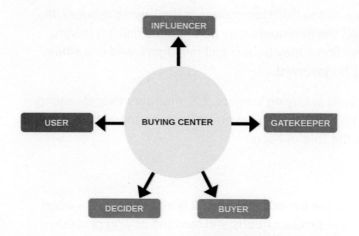

Responsible Marketing

Transparency and Communication in B2B Marketing

B2B marketing occurs when the customer is another business or organization. Business customers behave differently than traditional consumers, and are often engaged in significant purchase decisions that impact their ability to perform their own business or organizational functions.

We could explore a variety of responsible marketing challenges in the B2B context, but one of the most prevalent involves pricing. Because of the unique structure of B2B relationships, pricing can change by customer in a way that's different from consumers.

Consider the following fictional scenario:

Tom is a sales associate for a top robotic equipment manufacturing company that makes robotic arms for production lines for software companies. The product Tom sells is very specialized and there is little competition, yet a lot of demand, in the market.

One of Tom's main clients is Gary. Gary has been a customer for 15 years and has been a lucrative customer. Gary and Tom are also good friends. Gary tells Tom that his business suffered during COVID-19 and asked for a discount on a big order of 12 new robotic arms. Tom agreed, valuing the relationship with Gary, and offered a 20% discount off the normal asking price.

Phil is a potential new customer of Tom. Tom just read an article that Phil's firm just received $100 million in investment funding. Tom believes that Phil has a lot of money for the robotic arms and could be a great customer long term. In order to compensate for the lost revenue with Gary, Tom decides to charge Phil 20% over the normal asking price for the robotic arms. Phil, not knowing any better, agrees to the price.

Reflection Questions

1. Do you feel that Tom's behaviors are ethical or unethical? Explain your answer.

2. Should Tom charge both customers the same price?

Understanding Your Customer: Test

1. When customers look beyond their personal knowledge and experience to help them in buying something, they are engaging in

   A. ritual consumption.

   B. external information search.

   C. cognitive dissonance.

   D. business-to-business marketing.

   E. self-actualization.

2. Robin, a 28-year-old fashion designer, is very finicky about the brands that she consumes as a customer. She believes in buying brands that she has previously used. When Robin walks into Brown and Grey, a supermarket, she is most likely to engage in _____ to buy toothpaste and mouthwash.

   A. cognitive dissonance

   B. an evaluation of alternatives

   C. an internal information search

   D. impulse buying

   E. business-to-business marketing

3. Tom recently bought a pair of headphones, XOB 10, from Blue for $99. Within a month, Tom's friend, Leo, bought a brand new pair of headphones XGD 20, released by Soundz, online at a reduced price of $86. Tom tried the XGD 20 and found it to be significantly better the XOB 10. In this scenario, Tom is most likely to experience

   A. cognitive dissonance.

   B. self-actualization.

   C. problem recognition.

   D. external information search.

   E. criteria evaluation.

4. Consumer learning typically begins with

   A. knowing that a particular product exists.

   B. a response which attempts to satisfy a need or want.

   C. reinforcement that a particular product satisfies a need or want.

   D. cues that discourage the use of a particular product.

   E. a stimulus that encourages consumers to act to reduce a need or want.

5. According to Maslow's hierarchy of needs, once their physical needs have been satisfied, consumers' _____ take precedence.

   A. need for belonging

   B. safety needs

   C. esteem needs

   D. self-actualization needs

   E. need for food

6. Which of the following exemplifies impulse buying in a supermarket?

   A. Picking up a candy bar while waiting in line at the checkout counter

   B. Remembering to buy toothpaste while browsing the groceries sections

   C. Buying an reasonably priced watch from the accessories sections

   D. Picking up a bottle of wine to go with meat

   E. Opting for a different brand of tissues from the usual brand

7. Jerome has a dish full of dirty dishes and has run out of his favorite dishwashing liquid, so he heads to the corner market to pick up a new bottle. This type of consumer purchase exemplifies which variation of the consumer buying process?

   A. extended problem solving

   B. motivated purchase

   C. high-involvement purchase

   D. routine problem solving

   E. limited problem solving

8. The soles of Asha's Asics running shoes are wearing thin, so she needs to purchase a new pair. She knows the style she has worn the last two years has been updated, so she plans to purchase the same brand, but in the updated style. This type of purchase is considered a limited problem-solving purchase.

   A. TRUE

   B. FALSE

9. Hassan is getting ready to purchase his first home. His involvement throughout the buying process will likely be mid-level because this purchase is considered an extend problem-solving purchase.

   A. TRUE

   B. FALSE

10. In order to ensure that firms sell more products more profitably, marketers need only to understand *why* individuals and organizations make their purchases.

    A. TRUE

    B. FALSE

11. In which of the following would primary data be *most* helpful to a firm?

    A. determining whether price cuts on store-branded products impact what products consumers buy

    B. determining the competition's prices and inventory

    C. understanding the international market for its products

    D. identifying which in areas to open a second location

    E. determining interest rates before deciding whether to expand

12. A toy company is testing kites to see which one can be flown the highest. In this example, which of the following would be considered a dependent variable?

    A. the height the kite can be flown

    B. the temperature at the time the kite is flown

    C. the wind conditions at the time the kite is flown

    D. the type of kite being flown

    E. the person flying the kite

13. Antoine wants to know if his main rival in organic baked goods is increasing its prices. He talks to his rival's distributors, suppliers, and some of his rival's customers to get this information. Antoine is engaging in

    A. data mining.

    B. trending.

    C. user research.

    D. usage studies.

    E. competitive intelligence.

14. The marketing research industry relies on ethical standards to help gain the trust of consumers. Establishing trust

    A. increases individuals' willingness to participate in research.

    B. ensures accurate research results.

    C. informs companies when to cease research.

    D. decreases the cost of research.

    E. increases the cost of research.

15. The Blooms & Blossoms company has been selling its garden supplies to retail outlets for years. The company has decided it wants to tap into new markets, so it is looking to sell its garden supplies to county institutions for planting gardens in public parks. If the county institutions purchase the Blooms & Blossoms garden supplies, it would be an example of a(n)

    A. modified rebuy.

    B. first buy.

    C. new buy.

    D. straight rebuy.

    E. institutional buy.

16. Fenton orders the same office stationery each month for the same price from the same supplier. This is an example of a

   A. modified rebuy.

   B. quick buy.

   C. continual buy.

   D. new buy.

   E. straight rebuy.

17. A B2B seller can use mobile applications to create a competitive advantage by integrating its mobile technology into the its own supply chain.

   A. TRUE

   B. FALSE

18. E-commerce is *most* likely to be associated with which type of B2B buying situation?

   A. modified rebuy

   B. collaborative relationship

   C. new buy

   D. transactional relationship

   E. relationship marketing

19. This application is a type of enablement tool.

   A. delivery reminder

   B. CRM system

   C. e-commerce platform

   D. social media presence

   E. mobile application

20. Someone who is acting as an influencer in the buying center may be best illustrated by which of the following scenarios?

   A. The firm's purchasing agent has been asked to get the price of office supplies down before committing to the purchase.

   B. The company's CEO has said that the head engineer will choose what surveying equipment the company will buy.

   C. All information flow from potential suppliers is filtered through the executive assistant who makes appointments and distributes what he has received to different members of the buying center.

   D. The members of the call center have stated that the desks that are to be purchased for them must be adjustable so that they can have the option of standing.

   E. While discussing an upgrade to the company's computer software, the head of IT discussed current trends in computing that she would like the company to consider.

What To Expect

By the end of this lesson, you will be able to:

- Explain the purpose of accounting, as it relates to users of financial data and a company's success.
- Explain how a business's financial information is organized utilizing the six-step accounting process, in accordance with GAAP.

Chapter Topics

- **7-1** What Is Accounting?
- **7-2** Financial Statements: The Balance Sheet
- **7-3** Financial Statements: The Income Statement
- **7-4** Financial Statements: The Cash Flow Statement
- **7-5** Financial Ratios

Copyright © McGraw-Hill. Natee Meepian/Shutterstock

What Is Accounting?

Glitz n' Glamour: Naomi's Predicament

"Why Can't I Get Ahead?"

Naomi, a talented fashionista and small boutique owner, sighs heavily as she sinks into the chair across from her friend. Her friend, Luke, nods sympathetically. In the three years since she became an entrepreneur, Naomi's frustrations with her business, Glitz n' Glamour, have only grown.

In fact, it seems that every time they meet, she is struggling with the same mystery—she cannot understand why her seemingly "booming" business still isn't making a profit.

I just don't get it! Why am I not making any money? I work every day—even weekends!

Still? Wow. I mean, the store seems like business is booming! My wife said it was packed when she dropped in yesterday.

Yeah, things have really picked up in the last eight months, but when the bills come in, we always come up short. It's like all of those extra sales have had no impact at all. There must be something that I'm doing wrong.

Maybe it's marketing? Or, what about your financial records—are they accurate? I have a friend who is taking some business courses; would you like me to connect you?

Oh, would you? I'll take any help I can get.

Sure; why not? I'll make a call tonight. Let's meet back here tomorrow and I'll let you know what we discuss.

The next day, Naomi and Luke meet up to talk about his conversation with his friend.

Naomi, great news! My friend is happy to help you out. The best place to start is looking over your books with your accountant. You can set up a meeting, right? Or, if your accountant isn't available, I'm sure going through your books will be a good start. Naomi? Are you okay?

Whoa! Slow down! My books? My accountant? I don't have either of those. I mean, my business is really small, so I haven't really been managing it any differently than my personal spending. Is all of that really necessary?

Uh oh. It looks like Naomi is lacking key knowledge in the area of accounting. As a result, her business is underperforming, impacting her profits, and quite possibly jeopardizing the future success of Glitz n' Glamour.

In this lesson, we will work together to teach Naomi the fundamental accounting knowledge and skills she needs to implement appropriate accounting practices, build key financial statements, and analyze her business to make better (and more informed) business decisions and gain new insight on how to control her costs and increase her profits.

Accounting and Bookkeeping

Our first challenge is to help Naomi understand what accounting is, and why she should let us help her get her business on track using accounting functions.

Accounting measures, classifies, analyzes, and communicates financial information to people inside and outside a company.

Bookkeeping is one function of accounting, which includes recording a company's financial transactions.

Although many people (falsely) assume that all accountants are bookkeepers, accounting is more than just bookkeeping. Engaging in proper accounting functions can help business owners, employees, and outside parties—like suppliers or investors—make informed financial decisions.

Types and Functions of Accounting

Although it can seem daunting at first, accounting can be broken into two distinct types, with three different kinds of accounting professionals. Let's break it down.

Functions of Accounting

Managerial accounting is concerned with preparing accounting information and analyses for managers and other decision makers inside an organization.

Financial accounting is concerned with preparing accounting information and analyses primarily for people outside of the organization, such as stockholders, government agencies, creditors, lenders, suppliers, unions, customers, consumer groups, and so on.

Managerial Accounting	Financial Accounting
• Prepare accounting information and analyses for managers and decision makers.	• Prepare accounting information and analyses for people outside of the organization.
• Prepare and monitor department and company-wide budgets.	• Perform audits to ensure accuracy of the financial information.
• Analyze the costs of production and marketing.	• Prepare and file taxes and other documents for government agencies.
• Manage and control inventory and production costs.	

Managerial Accounting

Who uses managerial accounting?
Managers: use accounting info to make plans and guide the company to achieve goals and outcomes set by the business

Production managers: use sales forecasts to set production levels

Marketing managers: use accounting info to evaluate the impact of promotion strategies

Employees: use accounting info to evaluate the impact of promotion strategies

Financial Accounting

Who uses financial accounting?
Stockholders, investors: evaluate firm's financial health to see how well the firm is doing and whether it is profitable

Managers: plan, set goals, control

Employees: measure financial health

Lenders and suppliers: evaluate credit ratings

Government agencies: confirm taxes and regulatory compliance

Who Uses Accounting?

Types of Accountants

There are three types of accountants. **Private accountants** are internal accountants working for a single organization. On the other hand, **public accountants** provide accounting services to clients on a fee basis. Lastly, **not-for-profit accountants** work for governments and nonprofit organizations and perform the same services as for-profit accounts—except they're concerned primarily with efficiency, not profits. As entrepreneurs start their businesses, they may need to decide which one is their best fit.

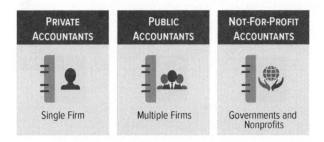

- Private accountants work for a single company or firm like Netflix.
- Public accountants work for many clients such as Glitz n' Glamour.
- Not-for-profit accountants specialize in nonprofit organizations such as the Red Cross.

Decisions

Naomi: Hi Luke!

Luke: Hey Naomi! What's up?

Naomi: Just overwhelmed, as usual.

Luke: Oh no, not again. Did you contact that _____ I told you about? You know, the accountant that is available to hire on a contract basis?

- **A.** certified public accountant (CPA)
- **B.** financial analyst
- **C.** work-for-hire
- **D.** private accountant

No, but I am going to this afternoon. Her name was Chitra, right? I'm hoping she can help, but what I really need is someone who will be able to look at my financials and tell me some options of what I need to do.

You're in luck, she specializes in _____, and she works with business owners to help them get their entire business under control from a numbers point of view.

- **A.** financial accoutning
- **B.** managerial accounting
- **C.** tax accounting
- **D.** CNA accoutning

Luke, you are totally getting my hopes up! I sure hope she can help me because I don't have the money to hire a(n) _____ to work here with me part-time, let alone full-time.

- **A.** public accountant
- **B.** in-house accountant
- **C.** private accountant
- **D.** broker

Yes, hiring on someone would cost quite a bit of money. I'm anxious to hear what you think of Chitra after your meeting today. Keep me posted!

Correct Answers: A; B; C

GAAP: The Trademark of a Reputable Accountant

Accounting relies on *standards*, known as **Generally Accepted Accounting Principles (GAAP)**, which ensure that financial statements are relevant, reliable, consistent, and comparable. When financial statements are consistent and comparable, stakeholders can compare them with earlier statements within the company and with statements from other companies.

RELEVANT
Information should help users understand the company's financial status and performance.

CONSISTENT
Information should always be based on the same assumptions and procedures. Any changes to this must be clearly explained.

RELIABLE
Information should be accurate, objective, and verifiable.

COMPARABLE
Information should allow users to compare data with other companies' data and with prior data from the company.

The Six-Step Accounting Process

To create *relevant*, *reliable*, *consistent*, and *comparable* financial statements, we must apply the **accounting process**, which involves six activities: collection, recording, classification, summarization, reporting, and analysis.

An adept accountant will:

1. Locate and sort records (collection).

2. Record daily transactions in journals (recording).

3. Organize journal entries in categories within a ledger (classification).

4. Test the accuracy of the ledger by running a trial balance (summarization).

5. Issue financial statements (reporting).

6. Assess the firm's financial condition via ratio analysis (analysis).

Collect: Find and Sort Records of Relevant Business Transactions

The accounting process begins by collecting the results of bookkeeping—that is, the records of all relevant business **transactions**: sales invoices, cash receipts, travel records, shipping documents, and so on. Today, many or most of these transactions exist as computerized data. The transactions are analyzed by the bookkeeper and sorted into meaningful categories according to various strategies, some of which involve ways of reducing tax obligations.

At Glitz n' Glamour, the business should be counting the revenue received from each sales transaction on a daily basis into an accounting system or spreadsheet. The company would also want to enter any data related to travel, shipping, or other transactions made to help run the business.

Record: Put Daily Transactions in Journals, Using Double-Entry Bookkeeping

The bookkeeper then records financial data from the original transaction documents in a journal, a record book, or part of a computer program containing the daily record of the firm's transactions, including a brief description of each.

At Glitz n' Glamour, the individual tasked with bookkeeping would want to ensure the transactions are accurately recorded into the firm's accounting system.

Double Entry Accounting System

The format for recording each journal entry is known as **double-entry bookkeeping** because each transaction is recorded in two different accounts to make sure each add up to the same amount, as a check on errors. Thus, if a bookkeeper enters $9.49 in one place, but mistakenly puts $9.94 in another, when the two accounts don't produce the same total it will be a tip-off that an error occurred.

With double-entry bookkeeping, notice that for each **debit**, there is an equal and opposite **credit**. Thus, the total of all debits must equal the total of all credits. If they don't, obviously an error has occurred. Recording each transaction as both a debit and a credit keeps the books in balance, as we'll discuss with the balance sheet.

Example: Singe-Entry Versus Double-Entry Bookkeeping

If you run a bicycle repair shop, a *single-entry bookkeeping transaction* to repair a tire would appear as follows:

Date	Service or Sale	Revenues	Expenses
February 6	Tire repair	$30.00	

A *double-entry bookkeeping transaction* would look like this:

Date	Service or Sale	Debit	Credit
February 6	Cash	$30.00	
	Revenue		$30.00

Classify: Put Journal Entries in Categories in a Ledger

Suppose you want to know what Naomi's **expenses** were for the past month. The **journal** will show these expenses scattered throughout, but not all lumped together. That is the purpose of the bookkeeper's transferring (usually on a monthly basis) journal entries to a **ledger**, a specialized record book or computer program that contains summaries of all journal transactions that are accumulated into specific categories. The ledger is divided into accounts, such as cash, inventories, and receivables.

At Glitz n' Glamour, the individual tasked with bookkeeping would want to ensure the transactions are accurately coded into the firms accounting system.

Summarize: Test the Accuracy of Ledger Data by Running a Trial Balance

At the end of every accounting period (every three months, for example), the bookkeeper does a check for accuracy. This is known as running a **trial balance**, making a summary of all the data in the ledgers to see if the figures are accurate, or balanced. Balanced means both columns in the double-entry format have similar totals—they balance each other.

At Glitz n' Glamour, the individual tasked with bookkeeping creates the trial balance making sure that each column in the double-entry format have similar totals.

Report: Issue Financial Statements Such as the Balance Sheet, Income Statement, and Statement of Cash Flows

Once the bookkeeper has the correct figures, he or she (or the accountant) can use that summarized data to issue three reports, or financial statements—the balance sheet, the income statement, and the statement of cash flows. Most companies prepare computer-generated financial statements every month, three months (quarterly), or six months (semiannually). Financial reports may also be tied to two other matters—the organization's fiscal year and the release of its annual report.

Naomi can use this information to run reports and share with key stakeholders such as managers, employees, suppliers, banks, accountants, or investors.

The Fiscal Year

Financial reports are required of all publicly traded companies at the end of the firm's **fiscal year**, the 12-month period designated for annual financial reporting purposes. This period may coincide with the end of:

- The calendar year, January 1 to December 31.
- The U.S. government's fiscal year, October 1 to September 30; for firms with government contracts, financial reports may appear sometime after September 30.
- A particular industry's natural cycle. In agriculture, for example, it is September 1 to August 31 (when the harvest is over).

The Annual Report

After the fiscal year-end, a firm issues an **annual report**, showing its financial condition and outlook for the future. This report is utilized by managers and executives to evaluate the health of the business as well as to plan for the next fiscal year. This information is also used by shareholders and potential investors to evaluate the organization's financial performance.

Analyze: Assess The Company's Financial Condition, Using Ratio Analysis

With the financial statements in hand, the accountant can then make an assessment of the company's financial condition.

Naomi can now analyze Glitz n' Glamour using verified data and validated accounting practices to better understand the performance of her business as well as the company's financial condition.

Decisions

Help Naomi align her accounting process. Choose the step in the process that best aligns with the statement.

1. Glitz n' Glamour should be counting the revenue received from each sales transaction each day into an accounting system or spreadsheet. The company should also enter any data related to shipping or other expenses made to help run the business.

    A. Summarize

    B. Analyze

    C. Record

    D. Collect

    E. Report

    F. Classify

2. The individual tasked with bookkeeping should ensure the transactions are accurately recorded and coded into the firms accounting system.

   A. Summarize

   B. Correct.

   C. Record

   D. Collect

   E. Classify

   F. Report

   G. Analyze

3. The individual responsible for bookkeeping would transfer journal entries into a ledger that shows the different account categories.

   A. Report

   B. Collect

   C. Summarize

   D. Analyze

   E. Record

   F. Classify

4. Naomi can use this information to run reports and share with key stakeholders such as managers, employees, suppliers, banks, accountants, or investors.

   A. Analyze

   B. Summarize

   C. Collect

   D. Classify

   E. Record

   F. Report

5. With the financial statements in hand, the accountant can then make an assessment of the company's financial condition. Naomi can now analyze Glitz n' Glamour using verified data and validated accounting practices to better understand the performance of her business as well as the company's financial condition.

   A. Collect

   B. Report

   C. Summarize

   D. Analyze

   E. Classify

   F. Record

Correct Answers: 1. D; 2. C; 3. F; 4. F; 5. D

Financial Statements: The Balance Sheet

By the end of this lesson, you will be able to:

- Using provided data, construct an accurate, well-organized balance sheet.

Naomi Studies the Numbers

"The Data Is in Disarray!"

Naomi has hired Chitra and is learning that her seemingly successful business isn't as healthy as it appears. She knows she needs some help understanding her financial statements and figuring out what to do.

Wow, I just looked at these financial reports. It's like reading a foreign language. The one thing I can tell is we have a lot less profit than I realized, and I am not sure how that happened. I know we have great products, awesome people, and we are implementing great processes. I want to use this information to grow our business. Do you think you can help me better understand these numbers and what they mean so I can better lead the company forward?

Definitely! Let's take a look at each of the financial statements. We will walk through them and then we'll look at some financial ratios to analyze the business.

Naomi seems very engaged and excited to learn more about her company's performance so she can better lead her company to growth and profitability. In this lesson, we will work together to teach her about one type of financial statement that is central to understanding her business: the balance sheet.

The Balance Sheet

Measuring the Health of a Business

Have you ever gone into the doctor for an annual check-up?

Healthy businesses use data to make informed decisions that will keep them healthy. In businesses, we utilize **financial statements** as metrics to analyze the health of the business. One of the primary statements used to measure a company's vitals is the **balance sheet**, which examines the overall health of the business.

As we begin to learn how to analyze the performance of a business, it is important to understand the key elements that make up the balance sheet.

The Accounting Equation

For a business to be successful it is mission critical for accountants, owners, and managers to understand the balance sheet. The **accounting equation** ensures that the balance sheet does exactly that—it balances. If applied correctly, total assets (things of value owned by a business) should equal total liabilities (what the business owes) plus owners' equity. Accountants employ the accounting equation to determine what a firm is worth.

The accounting equation is:

Reviewing a Balance Sheet

The balance sheet reports a firm's financial condition at a given time by showing its **assets**, **liabilities**, and **owners' equity**. It is called a balance sheet because it shows a balance between a firm's (a) assets and (b) liabilities plus owners' equity.

Although Naomi's accountant has prepared a balance sheet for Glitz n' Glamour, Naomi isn't sure how to read it. Let's use it to walk her through the basics and break down each section. Starting with the balance sheet in its totality.

You'll see it is laid out starting with the company's assets, which are the items the business has that are of value. Then you'll see the company's liabilities, what they owe to others, and lastly you'll see the company's owners' equity, which is reflective of both the owners' investments into the business, any **stock** sold to investors, and any earnings (profits) retained by the company.

The Balance Sheet

Glitz n' Glamour

Balance Sheet

	FY-2018	FY-2019
Current Assets		
Cash	$ 9,000.00	$ 15,500.00
Accounts Receivable	53,000.00	45,000.00
Inventory	80,000.00	76,000.00
Prepaid Expenses	9,500.00	8,900.00
Total Current Assets	**151,500.00**	**145,400.00**
Fixed Assets		
Land	100,000.00	100,000.00
Building	180,000.00	180,000.00
Less: Accumulated Depreciation	(2,000.00)	(4,000.00)
Equipment	47,000.00	50,000.00
Less: Accumulated Depreciation	(8,000.00)	(10,000.00)
Furniture and Fixtures	36,000.00	45,000.00
Less: Accumulated Depreciation	(6,000.00)	(7,000.00)
Total Fixed Assets	**347,000.00**	**354,000.00**
Intangible Assets		
Goodwill	23,000.00	19,600.00
Total Assets	**521,500.00**	**519,000.00**
Current Liabilities		
Accounts Payable	40,000.00	45,000.00
Current Portion Notes Payable	40,000.00	40,000.00
Total Current Liabilities	**80,000.00**	**85,000.00**
Long-term Liabilities		
Notes Payable	120,000.00	100,000.00
Mortgage Payable	280,000.00	270,000.00
Total Long-term Liabilities	**400,000.00**	**370,000.00**
Stockholders' Equity		
Common Stock	37,000.00	37,000.00
Retained Earnings	4,500.00	27,000.00
Total Stockholders' Equity	**41,500.00**	**64,000.00**
Total Liabilities and Stockholders' Equity	**521,500.00**	**519,000.00**

Let's dissect the balance sheet based on its component parts.

Assets: What Things of Value Do We Own?

A balance sheet always begins with the items the firm owns. This includes buildings, land, supplies, inventories, cash, money owed to the firm, patents, and trademarks.

The assets are then broken into current, fixed, and intangible.

Current Assets: Items That Can Be Converted to Cash within One Year

Current assets are defined as items that can be converted into cash within one year. This includes not only cash itself (currency and coin), but also marketable securities, accounts receivable, and merchandise inventory.

The essential characteristic of current assets is their **liquidity**. Current assets are easily converted into cash, whereas assets such as land and buildings are not. Besides cash, two common items found on most balance sheets include accounts receivable and inventory.

- **Accounts receivable** is the total amount owed to a firm from customers who have purchased goods or services on credit. Accounts receivable is often referred to as A/R.
- **Inventory** on the balance sheet refers to the value of the products or merchandise that is being held for resale to customers.

Additionally, some businesses may also show prepaid expenses and marketable securities as part of their current assets.

- **Marketable securities** are stocks, bonds, government securities, and money market certificates, which can be easily converted to cash.
- **Prepaid expenses** are expenditures that are paid for in one accounting period, even though assets will not be consumed until a later time. An example is insurance.

For Glitz n' Glamour we see that Naomi's business has current assets that include cash, accounts receivable, and inventory. Naomi has prepaid some of her expenses as well. These represent the current assets of the company.

Current Assets on the Balance Sheet

Glitz n' Glamour

Balance Sheet Breakout

	FY-2018	FY-2019
Current Assets		
Cash	$ 9,000	$ 15,500
Accounts Receivable	53,000	45,000
Inventory	80,000	76,000
Prepaid Expenses	9,500	8,900
Total Current Assets	**151,500**	**145,400**

Fixed Assets: Items Held for a Longer Time

Often called *property, plant, and equipment*, **fixed assets** are items that are held for a long time and are relatively permanent, such as land, buildings and improvements, equipment and vehicles, and furniture and fixtures. Fixed assets are expected to be used for several years.

Accumulated depreciation is the reduction in value of assets to reflect their wearing down or obsolescence over time.

For Glitz n' Glamour you'll see that Naomi's accountant has accounted for land, building, equipment, furniture, and fixtures, and then reduced the value by applying the accumulated depreciation.

Fixed Assets on the Balance Sheet

Glitz n' Glamour

Balance Sheet Breakout

	FY-2018		FY-2019	
Fixed Assets				
Land		100,000		100,000
Building	180,000		180,000	
Less: Accumulated Depreciation	(2,000)	178,000	(4,000)	176,000
Equipment	47,000		50,000	
Less: Accumulated Depreciation	(8,000)	39,000	(10,000)	40,000
Furniture and Fixtures	36,000		45,000	
Less: Accumulated Depreciation	(6,000)	30,000	(7,000)	38,000
Total Fixed Assets		**347,000**		**354,000**

Intangible Assets: Valuable Assets That Aren't Physical Objects

Intangible assets are assets that are not physical objects but are nonetheless valuable, such as patents, trademarks, and goodwill. **Goodwill** is an amount paid for a business beyond the value of its other assets, based on its reputation, customer list, loyal employees, and similar intangibles.

Naomi believes the Glitz n' Glamour brand is worth the amount noted on her balance sheet based on its reputation and the customer list she created over the years.

Intangible Assets on the Balance Sheet

Glitz n' Glamour

Balance Sheet Breakout

	FY-2018	FY-2019
Intangible Assets		
Goodwill	23,000	19,600

Understanding Liabilities: What Are Our Debts to Outsiders?

A **liability** is a debt owed by a firm to an outside individual or organization. Examples: Vendors may deliver supplies but not insist on immediate payment, allowing the firms 30 or 60 days to pay their bills. Banks may loan farmers money to enable them to operate, but want to be paid after the crop has been harvested and sold. Employees who have provided labor but not yet been paid also represent a debt to the firm.

Liabilities are of two types: current and long term.

Current Liabilities: Payments Due within One Year or Less

Current liabilities are obligations in which payments are due within one year or less. The most common current liabilities are accounts payable and notes payable.

- **Accounts payable** is money owed to others that the firm has not yet paid. If the company has not yet paid its electricity bill this month, that debt belongs in accounts payable. Accounts payable is commonly referred to as A/P.

- **Notes payable** is money owed on a loan based on a promise (either short term or long term) the firm made. If you arranged a bank loan, for example, you would be obligated to pay it back by a prearranged date.

Glitz n' Glamour currently has debts that it owes to vendors that we can see on there A/P. The company is also paying back some loans, which we can see through their notes payable.

Current Liabilities on the Balance Sheet

Glitz n' Glamour

Balance **Sheet Breakout**

	FY-2018	FY-2019
Current Liabilities		
Accounts Payable	40,000	45,000
Current Portion Notes Payable	40,000	40,000
Total Current Liabilities	**80,000**	**85,000**

Long-Term Liabilities: Payments Due in One Year or More

Long-term liabilities are obligations in which payments are due in one year or more, such as a long-term loan from a bank or insurance company.

- Two common long-term liabilities are **notes payable** and **bonds payable**, long-term liabilities that represent money lent to the firm that must be paid off.

- In some cases, a business may also own property, and as a result, have a **mortgage** it must repay. Because those payments are structured over a long period of time, they are categorized as long-term liabilities.

Glitz n' Glamour has a long-term loan they are paying back. In addition, the business owns a building, which is why it has a mortgage listed under the long-term liabilities section.

Long-Term Liabilities on the Balance Sheet

Glitz n' Glamour

Balance **Sheet Breakout**

	FY-2018	FY-2019
Long-Term Liabilities		
Notes Payable	120,000	100,000
Mortgage Payable	280,000	270,000
Total Long-Term Liabilities	**400,000**	**370,000**

Owners' Equity: "What Is Our Value If We Were to Sell Our Assets and Pay Off Our Debts?"

Owners' equity, or stockholders' equity, represents the value of a firm if its assets were sold and its debts paid. Owners' equity is considered important because it is used to indicate a company's financial strength and stability. Before making loans to a company, for example, lenders want to know the amount of owners' equity in it.

Owners' equity consists of (1) common stock and (2) retained earnings.

- **Common stock** refers to ownership in a business. As it pertains to the balance sheet, common stock accounts for the amount of stock that has been allocated to owners of the company.

- **Retained earnings** are net profits minus **dividend** payments made to stockholders; that is, they are earnings retained by a firm for its own use—for buying more land and buildings, say, or acquiring other companies.

At Glitz n' Glamour, Naomi made the business decision to issue common stock to some investors, meaning she allowed individuals to invest money in her company, and as result they received equity through stock in the company. The value of the investment is shown on the balance sheet as common stock. She has also been keeping some of the annual profit, which appears as retained earnings for Glitz n' Glamour.

Owners' Equity on the Balance Sheet

Glitz n' Glamour

Balance **Sheet Breakout**

	FY-2018	FY-2019
Owners' Equity		
Common Stock	37,000	37,000
Retained Earnings	4,500	27,000
Total Owners' Equity	**41,500**	**64,000**

Decisions

Chitra, I sure am glad I hired you on as my CPA. I am lost when it comes to all this financial information.

No problem, and believe it or not, you will be making sense of this stuff in no time. It's probably best if we start by trying to figure out your net worth by looking at what you have and what you owe on the _____.

A. asset statement

B. balance sheet

C. income statement

D. statement of cash flows

Net worth? I'm sure it's negative by now!

Don't be too sure about that! We can find out pretty quickly if you're wrong. All we need to do is look at how your assets balance out against your liabilities. This is called the _____.

A. asset paradox

B. Pythagorean theorem

C. fundamental accounting equation

D. balance sheet equation

Alright, so if we assess what I owe, are we looking only at the _____? You know, the stuff I owe to people in the long term?

A. short-term liabilities

B. owners' equity

C. long-term liabilities

D. total liabilities

No, we will look at all your liabilities, and then we will compare that to your asset value. Once we do this, we will know your _____, and that will give us a good idea of where you stand currently.

A. liquidity

B. notes payable

C. retained earnings

D. owners' equity

Yes, I see what you are talking about now. At least this will give me a firm footing of how Glitz n' Glamour is doing right now! It's better to know than not to know. Let's see where we stand!

Correct Answers: B; C; C; D

Lesson 7-3

Financial Statements: The Income Statement

By the end of this lesson, you will be able to:

- Using provided data, construct an accurate, well-organized income statement.

The Bottom Line: Profit and Loss

"Are We Making Money?"

Naomi is excited to learn more about financial statements and is anxious to better understand the income statement, so she meets with Chitra once more.

I just ran a report out of our software from last year's numbers: cost of goods sold, gross profit, gross sales, net profit, and net sales. I know there is a difference, and if I can get a hang of what's on this report, I can help our business grow and thrive! Do you think you can walk me through the income statement to help me better understand our numbers?

I know these terms seem a bit abstract, but once we go through them I think you'll be able to better understand how this statement is a reflection of all the hard work you and your team do, and it will give you an opportunity to see where you may want to make some adjustment to your business strategy.

Naomi is excited to continue her growth as a business owner. In this lesson, we'll analyze whether a business is making money by learning about the income statement.

The Income Statement

The **income statement**, once known as the profit-and-loss statement, shows a firm's revenues and expenses for a particular time period and the resulting profit or loss. The income statement itself has four principal parts: (1) sales revenue, (2) cost of goods sold, (3) operating expenses, all of which lead to the bottom line of (4) net income.

The Income Statement

Glitz n' Glamour

Income Statement FY 2019

Sales		871,000
Less: Sales Returns and Allowances		(11,000)
Net Sales		860,000
Less: Cost of Goods Sold		(516,000)
Gross Profit		344,000
Less: Operating Expenses		
Salaries and Wages	187,000	
Advertising	42,500	
Insurance	28,000	
Property Taxes	17,700	
Utilities	15,000	
Interest	14,500	
Depreciation/Amortization	8,400	
Supplies	6,200	
Total Operating Expenses		(319,300)
Income before Taxes		24,700
Less: Income Tax Expense		(2,200)
Net Income		22,500

Gross Profit: Net Sales and Our Cost of Goods

Organizations first must account for their overall revenue, and then they must take into account any returns and allowances, then they must account for what it costs to acquire, make, or provide merchandise. As a result, a company can determine their gross profits.

This portion of the income statement is comprised of the following:

- **Sales revenues** are the funds received from the sales of goods and services during a certain period. The figure of interest to us is arrived at by taking gross sales and subtracting sales returns and allowances, to arrive at net sales.

- **Gross sales** are the funds received from all sales of the firm's products.

- **Sales returns** are products that customers return to the company for a refund.

- **Allowances** are partial refunds to customers for damaged products they choose to keep, not return.

- **Net sales** refers to the money resulting after sales returns and allowances are subtracted from gross sales.

- **Cost of goods sold** is commonly referred to as COGS and represents the cost of producing a firm's merchandise for sale during a certain period. For a company like Naomi's, this may include purchasing garments from manufacturers and designers.

For Glitz n' Glamour, Naomi's company was able to generate $871,000 in sales, she then deducts $11,000 for returns and allowances, which left her with net sales of $860,000. Then the company must subtract out the cost of goods sold, which leaves the company with $344,000. This amount refers to the **gross profit**, or gross margin, which is the amount remaining after the cost of goods sold is subtracted from the net sales.

Gross Profit on the Income Statement

Glitz n' Glamour
Income Statement FY 2019 Breakout

Sales	871,000
Less: Sales Returns and Allowances	(11,000)
Net Sales	860,000
Less: Cost of Goods Sold	(516,000)
Gross Profit	344,000

Operating Expenses: "What Were Our Selling and Administrative Expenses?"

Although a business may have a strong gross profit, the firm must still account for the expenses associated with the business. We refer to these as **operating expenses**, which are selling and administrative expenses incurred by the company.

- **Selling expenses** are all the expenses incurred in marketing the firm's products, such as salespeople's salaries, advertising, and supplies.

- **Administrative expenses** are costs incurred for the general operation of the business, such as salaries, supplies, depreciation, insurance, rent, and utilities.

For Glitz n' Glamour, the company has total operating expenses of $319,300. Its selling expenses are related to salaries and wages as well as advertising. The company's administrative expenses may also include salaries and wages as the company may include wages for all employees in one line item. It also includes insurance, property taxes, interest, utilities, depreciation/amortization, and supplies.

Operating Expenses on the Income Statement

Glitz n' Glamour
Income Statement FY 2019 Breakout

Less: Operating Expenses		
Salaries and Wages	187,000	
Advertising	42,500	
Insurance	28,000	
Property Taxes	17,700	
Utilities	15,000	
Interest	14,500	
Depreciation/Amortization	8,400	
Supplies	6,200	
Total Operating Expenses		(319,300)

Now that we know the companies gross profit and operating expenses we can determine the companies profit before and after taxes.

Net Income—The Bottom Line: "What After-Tax Profit or Loss Did We End Up With?"

Net income, the firm's profit or loss after paying income taxes, is determined by subtracting expenses from revenues. Net income is an important measure of company success or failure.

In the case of Glitz n' Glamour, Naomi's business made $24,700, which is the amount the government uses to calculate what the business owes in taxes. Therefore, after Glitz n' Glamour pays $2,200 in taxes, the company made $22,500. Because this is the last line of the income statement and tells whether or not a company made or lost money, it is often called the bottom line.

Net Income on the Income Statement

Glitz n' Glamour

Income Statement FY 2019 Breakout

Income before Taxes	24,700
Less: Income Tax Expense	(2,200)
Net Income	22,500

Decisions

Chitra, I hate to take out my frustrations on you each time we meet, but I just don't know what in the world is going on with my business! I am selling so much inventory, but I really don't know why I am not making much money in the end.

That's alright, Naomi, I have many clients who do all sorts of things in my office: they cry, yell, laugh hysterically, and even just stare at the wall. Sometimes numbers overwhelm people, but when you mention selling lots of product and not having as much money as you thought you'd have, we need to start by looking at the _____.

    A. balance sheet

    B. sales forecast

    C. income statement

    D. inventory turnover statement

Well I had $871,000 in sales revenue, so where did it all go?

In order to get to the bottom line, we need to check out where your money is going. Your net income before taxes is calculated by _____.

A. subtracting gross profit from sales revenue

B. subtracting net income before taxes from gross profit

C. adding expenses to net income before taxes

D. subtracting expenses from gross profit

Yes, and I am sitting at $24,700. This seems way too low. What is happening?

Your _____ is/are way too high.

A. expenses

B. net income after taxes

C. sales revenue

D. gross profit

I see it now, and I really wondered about this. I need to start cutting back, right?

Well, that is probably true. The two places you are spending money before taxes would be _____ and _____.

A. expenses; net income before taxes

B. cost of goods sold; expenses

C. expenses; gross profit

D. sales revenue; expenses

Yes, and I need to get those under control. See, I am feeling empowered already!

Correct Answers: C; D; A; B

Financial Statements: The Cash Flow Statement

By the end of this lesson, you will be able to:

- Using provided data, construct an accurate, well-organized cash flow statement.

The Cash Flow Statement: How Money Came and Went

"Where Is the Cash Flowing?"

Naomi is gaining a better understanding of her financial statements, she is curious about what she can learn about a company's operations by viewing a statement of cash flows. She follows up with Chitra for more information.

Okay, so one statement tells me the big picture of the business, another tells me how I am doing in relation to sales and expenses; I wish there was a way to dial in how money was coming and going from the business.

We do have a financial statement that helps us do just that.

You've got to be kidding me!

Hey, I wouldn't joke when it comes to your cash! The statement is called the cash flow statement, and it tells us how money is coming into the business and where it is going. Let me show you!

Naomi can hardly contain her excitement! With this newfound knowledge, she is gaining confidence in her ability to better run Glitz n' Glamour. In this lesson, you will learn how to find the cash coming in and out of a business.

The Cash Flow Statement

The **cash flow statement**, or statement of cash flows, reports over a period of time, first, the firm's cash receipts and, second, disbursement related to the firm's (1) operating, (2) investing, and (3) financing activities, which leads to the bottom line of (4) the cash balance.

Cash Flow Statement

Glitz n' Glamour

Cash Flow Statement FY 2019

Cash from Operating Activities		
Cash Received from Sales	868,000	
Cash Paid for Inventory	(507,000)	
Cash Paid for Operating Expenses	(312,500)	
Net Cash Provided by Operating Activities		**48,500**
Cash from Investing Activities		
Purchase of Equipment	(3000)	
Purchase of Furniture and Fixtures	(9,000)	
Net Cash Used by Investing Activities		**(12,000)**
Cash from Financing Activities		
Payment of Notes Payable	(20,000)	
Payment of Mortgage Payable	(10,000)	
Net Cash Used by Financing Activities		**(30,000)**
Net Increase in Cash		**6,500**
Cash Balance FY-2018		**9,000**
Cash Balance FY-2019		**15,500**

Operating Activities: "What Were the Incomes and Costs of Running Our Business?"

Cash flows from operating activities reflect income from sales and other income and payments for salaries, interest, taxes, and so forth. These are the ordinary costs of running a business.

For Glitz n' Glamour we can see the cash coming from sales and going out via purchasing inventory and paying operating expenses.

Glitz n' Glamour

Statement of Cash Flows FY 2019 Breakout

Cash from Operating Activities

Cash Received from Sales	868,000
Cash Paid for Inventory	(507,000)
Cash Paid for Operating Expenses	(312,500)
Net Cash Provided by Operating Activities	**48,500**

Investing Activities: "How Much Did We Earn from Our Investments?"

Cash flows from investing activities reflect the cash received from selling long-term assets, cash spent on buying equipment, and other investment activities.

Glitz n' Glamour did not earn any money from investing. Although they did make some investments in equipment, furniture, and fixtures.

Glitz n' Glamour

Statement of Cash Flows FY 2019 Breakout

Cash from Investing Activities

Purchase of Equipment	(3,000)
Purchase of Furniture and Fixtures	(9,000)
Net Cash Used by Investing Activities	**(12,000)**

Financing Activities: "How Well Did We Do from Our Loans to Others, Payment of Debt, and Our Stock Transactions?"

Cash flows from financing activities reflect the inflows and outflows of borrowed funds and long-term debt, sales of new stock, and payment of dividends.

Glitz n' Glamour is continuing to pay their loan via the notes payable and the building they own through the mortgage.

Glitz n' Glamour

Statement of Cash Flows FY 2019 Breakout

Cash from Financing Activities

Payment of Notes Payable	(20,000)
Payment of Mortgage Payable	(10,000)
Net Cash Used by Financing Activities	**(30,000)**

Cash Balance: "How Much Money Is Available at the End of the Year?"

Cash balance is the balance in the firm's cash account at the end of the year. Analyzing and understanding cash flow is vital to the success of any firm because many failed businesses blame their financial distress on inadequate cash flow.

Over the year Naomi's business has seen a cash increase of $6,500, and for the latest fiscal year, Glitz n' Glamour has a cash balance of $15,500.

Glitz n' Glamour

Statement of Cash Flows FY 2019 Breakout

Net Increase in Cash	6,500
Cash Balance FY-2018	9,000
Cash Balance FY-2019	15,500

Decisions

Chitra, is there a statement that would explain how much money I actually have right now?

Yes, it's the statement of cash flows, and it consists of _____.

A. operating cash, financing activities, expenses, and your cash balance

B. financing activities, sales revenues, expenses, net income after taxes, and your bottom line

C. sales revenues, expenses, and your cash balance

D. operating cash, investing cash, financing activity, and your cash balance.

Great, because I need to know how much money I can spend this month, and with all this money up in the air, I am hoping my checks don't bounce at the bank! Haha!

Yes, we can see these numbers because we can create a cash flow statement _____.

A. from the net income after tax numbers from the income statement

B. for any time period that we want

C. only monthly

D. only annually

If I'm looking at this correctly, I have more money in the end this year than I did last year, and this is called a _____.

A. profit

B. bottom line

C. net increase in cash

D. net income in cash

That's right, so keep it up, and next year might even be better!

Correct Answers: D; B; C

Lesson 7-5

Financial Ratios

By the end of this lesson, you will be able to:

- Interpret financial ratios in relation to organizational performance.

Using Financial Data to Analyze the Business

"What Do We Do with the Data?"

Naomi, elated with financial statements in hand, started to ponder how her business compares to others in her industry. She also wondered if there were any goals, standards, or benchmarks she should be comparing her business against. These thoughts had her hungry to learn more about accounting.

Having these financial statements is really helpful in understanding the business. I guess I wish there was a way to determine how well we are doing.

Naomi, we absolutely have a way to accomplish this goal.

Well, I know we have the accounting equation, profit and loss, and cash balance, but I'd like to start to analyze what we can do to make the business perform better.

I can help you with that. Now that we have the financial statements, we can conduct financial analysis using different ratios. These numbers can tell you how the business is doing and give you a chance to set some goals.

Naomi is ready to gain a deeper understanding of these numbers and how they can help her set new goals for the future of the business.

In this lesson, we will work together to teach Naomi how she can utilize the data from her financial statements to analyze her business's performance by using ratio analysis. Once Naomi can benchmark her firm's performance and become more aware of how her company operates, she will be more informed and empowered to make business decisions for the future.

Introduction to Accounting Ratios

Accountants don't just prepare financial statements. They also analyze the financial data to provide the firm's managers and investors with a better understanding of financial performance. One way of doing so is through ratio analysis. (Two other ways are horizontal analysis and vertical analysis, which can be used to compare financial statement numbers and ratios over a number of years. Expect to learn more about these analysis types in later studies.)

Ratio analysis uses one of a number of financial ratios—such as liquidity, efficiency, leverage, and profitability—to evaluate variables in a financial statement. Ratio analysis can be used to analyze a firm's performance (1) compared to its stated objectives and (2) compared to the performance of similar firms.

Ratio analysis evaluates the variables within a financial statement. Four types of financial ratios are:

1. **Liquidity ratios**, which determine how well a firm can pay its liabilities as they come due

2. **Activity ratios**, which determine how well the firm manages its assets to generate revenue

3. **Debt to owners' equity ratios**, which determine how much the firm relies on borrowing to finance its operations

4. **Profitability ratios**, which determine how high the firm's profits are in relation to its sales, assets, or owners' equity

Understanding Liquidity Ratios

We discuss four types of financial ratios in this lesson starting with the liquidity ratios.

Liquidity Ratios: Current Ratio and Acid-Test Ratio

Liquidity ratios measure a firm's ability to meet its short-term obligations when they become due. They are of interest to anyone who wants to know whether a firm is able to pay its short-term debts on time.

Two important liquidity ratios are:

- The **current ratio** consists of current assets divided by current liabilities. A goal for most businesses is to have 2.0 or higher ratio. The reason this is the benchmark is because the company has a two to one ratio of assets to liabilities, meaning it could more than likely pay back its current debts.

- The **acid-test ratio**, or quick ratio, consists of cash + marketable securities + receivables, all divided by current liabilities. The standard for this ratio is 1.0 or higher. This benchmark looks at a company's most liquid assets compared to its current debt. Therefore, a company which has a ratio of $1.00 of assets to $1.00 of liabilities has a strong likelihood of being able to service their current debts.

Current Ratio Equation Applied at Glitz n' Glamour

For Glitz n' Glamour, the balance sheet for fiscal year 2018 indicates that current assets are $145,400 and current liabilities are $85,000. When the assets are divided by the liabilities it produces a ratio of 1.71. This means the company has $1.71 in assets for every $1.00 in liabilities. Although this indicates there are enough current assets to manage the current liabilities, most banks and investors feel more secure lending and investing in a company that has a current ratio of 2.0 ($2.00 of assets for every $1.00 of liabilities) or higher.

$$current\ ratio = \frac{current\ assets}{current\ liabilities} = \frac{\$145,400}{\$85,000} = 1.71$$

Acid-Test Ratio Equation Applied at Glitz n' Glamour

Glitz n' Glamour's balance sheet for fiscal year 2018 shows $15,500 in cash, $0.00 in marketable securities, and $45,000 in accounts receivables, totaling $60,500. The current liabilities are $85,000.

$$acid\text{-}test\ ratio = \frac{cash + marketable\ securities + receivables}{current\ liabilities} = \frac{\$15,500 + \$0.00 + \$45,000}{\$85,000}$$

With a ratio of 0.71, Glitz n' Glamour is below the conventional acid-test ratio of 1.0. Because the result is less than 1.0, Naomi might have to boost her cash by borrowing from a high-cost lender, obtaining additional cash from investors, reducing dividend payments to stockholders, or trying to sell inventory.

Decisions

Hi, Chitra!

Naomi! How is Glitz n' Glamour going lately?

It's picking up this season, and I even thought about trying to buy a new inventory tagging system so I can keep track of my inventory in real-time. Problem is, I decided to try and take out a short-term loan of $7,500 from the bank to buy the tagging system, and the loan officer kept asking for my financial ratios. I didn't know what to tell him.

Yes, financial ratios. That's music to my ears! Well, almost!

I found it to be more like nails on a chalkboard.

Haha! All the information you need to get these financial ratios is already at your fingertips in your financial statements. I bet the loan officer wanted your current ratio, right? It compares _____.

   A. total assets with total liabilities

   B. current assets with total liabilities

   C. current liabilities with acid assets

   D. current assets with current liabilities

Yes, he did, and I asked why, and he said it would give him _____.

   A. a better idea if Glitz n' Glamour could turn its inventory into cash quickly

   B. a better idea if Glitz n' Glamour could have cash left over after the loan is made

   C. a better idea if Glitz n' Glamour could service its current debts

   D. a better idea if Glitz n' Glamour could pay back its current debts

Indeed, that is true. Did he also ask about the _____ ratio? It indicates if the company is able to service its immediate debts, thereby giving the bank an indication of how likely you would be in making your loan payments each month.

A. ending cash

B. debt

C. acid-test

D. current

Yes, he did, but I looked a bit like a deer in the headlights by then, so I gracefully bowed out of his office until I could meet with you. I guess I didn't realize these ratios were just figures derived from information I already had. Well, I guess I'm learning day by day.

Correct Answers: D; C

Understanding Activity Ratios

Activity or efficiency ratios are used to evaluate how well management uses a firm's assets to generate revenue. That is, they express how well a company can turn its assets (such as inventory) into cash to pay its short-term debts.

One of the primary ratios a company uses in determining its efficiency is the inventory turnover ratio, which measures the number of times a company sells its inventory in a year's time. The more frequently a firm can sell its inventory, the greater its revenue.

Inventory Turnover Ratio

The **inventory turnover ratio** consists of cost of goods sold in one year divided by the average value of the inventory.

Inventory turnover ratios vary by industry and by company, with Nike, for example, turning over its inventory about 4.5 times a year and Starbucks 14.7 times a year. A high ratio may indicate efficiency. A low ratio may indicate too much obsolete inventory or a need to sharpen the buying strategy. What is considered a "good" number of turns varies between industries.

Inventory Turnover Ratio Applied at Glitz n' Glamour

Glitz n' Glamour's income statement shows costs of goods sold was $516,000. Although for simplicity's sake we did not show the average inventory on that statement, it is determined by adding the inventory value at the end of fiscal year 2018 to the inventory value at the end of the previous year (fiscal year 2017) and dividing by 2, which is $78,000.

$$inventory\ turnover\ ratio = \frac{cost\ of\ goods\ sold}{average\ inventory} = \frac{\$516,500}{\$78,000} = 6.62$$

Glitz n' Glamour's inventory turnover ratio is somewhat low. It may have too much obsolete inventory or a buying strategy that needs to be sharpened.

Decisions

Hey Naomi! I just was in the neighborhood, and I thought I'd pop in to say hello!

Luke! So glad you stopped in! Check out these we just got in!

Sweet! These shirts are wicked cool!

Yes, they are fab! I sure hope they sell better than the ones over in that corner of the shop. Those even have a bit of dust on them! Dust, man! Hard to believe I'm actually dusting my inventory!

No kidding! That's not a good sign. Your _____ must be pretty low.

    A. sales to no-sales ratio

    B. asset sales ratio

    C. sales margins

    D. inventory turnover

You know about that? Wow, why have I been spending all this money on my accountant! Ha!

No, not really, but I guess I got lucky with my wording! Seriously, I do recall learning about it, but how does it work again?

It _____ in order to find out how frequently I'm selling off my current inventory of products.

A. analyzes the inventory turnover divided by the industry average turnover

B. analyzes the cost of goods sold in one year divided by the average value of the inventory

C. analyzes the cost of goods sold in one year divided by sales revenue

Interesting. So, how do you know if you are on track then?

Most companies compare their figures to _____.

A. the best company in their industry

B. the firm's figures for the previous year

C. industry averages

D. the worst company in their industry

So how is yours then?

It's a bit low, and as you can tell from the dust, I need to get crackin' on selling those clothes!

Correct Answers: D; B; C

Understanding Debt to Owners' Equity Ratios

The **debt to owners' equity ratio** is a measure of the extent to which a company uses debt, such as bank loans, to finance its operations. It is found by dividing total liabilities by owners' equity. If a firm takes on too much debt, it may have problems repaying borrowed funds or paying dividends to stockholders.

Debt to Owners' Equity Ratio Applied at Glitz n' Glamour

Glitz n' Glamour's balance sheet shows that total liabilities were $455,000 and total owners' equity was $64,000.

$$\text{debt to owners' equity ratio} = \frac{\text{total liabilities}}{\text{owners' equity}} = \frac{\$455,000}{\$64,000} = 711\%$$

With a ratio of 711%, this means that Glitz n' Glamour has borrowed $711 for every $100 the owners have provided. Clearly, Glitz n' Glamour has a high debt to equity ratio. However, it is important to compare their debt ratio to other firms in the their industry. With a a debt ratio over 100%, it may be more difficult for the company to borrow money from lenders. For some lenders and investors, a 100% ratio is acceptable; however, more conservative investors and lenders like to see this ratio around 70% or lower. Glitz n' Glamour far exceeds the preferred ratio, and the owner should look further into company's debt.

Decisions

Hi Naomi! I got your message this morning on my voicemail, and I was close by so I decided to just drop by and chat.

Thanks for coming by, Chitra! I had some questions about my debt ratio. The loan officer at the bank needs this ratio before considering my loan request, but I didn't know what to tell him. Where do I find this?

Sure, no problem. The debt to equity ratio is calculated by _____.

    A. dividing current liabilities to owners' equity

    B. multiplying owners' equity by total liabilities

    C. dividing total liabilities by owners' equity

    D. dividing total equity by total liabilities

Okay, but what is that really showing me?

Good question. The debt to equity ratio is most commonly used to _____.

    A. show what extent a business uses debt to finance its operations

    B. show how much debt the business has on its balance sheet

    C. show how much the business is worth through its equity

    D. show how much profit a business has earned

That makes sense now. The calculation is pretty straight forward, and I think Glitz n' Glamour needs to re-think taking on any new debt until we can get that ratio down.

True. Taking on debt is always a big decision, so think hard about it, but you also need to think hard if you can live without that real-time inventory tagging system. That system really might free up some of your time to devote to other things.

Yeah, like trying to pick some better shirts that will sell fast!

Correct Answers: C; A

Understanding Profitability Ratios

Profitability ratios are used to measure how well profits are doing in relation to the firm's sales, assets, or owners' equity. Three important profitability ratios are return on sales (profit margin), return on assets, and return on owners' equity.

Glitz n' Glamour may know what its net income is after taxes—$22,500, according to the income statement. However, to determine how well the firm is using its resources, Naomi will want to put this result into each of the three ratios to get a better understanding of her companies performance.

Return on Sales

The **return on sales**, or profit margin, is net income divided by sales. This information helps the company understand how much profit is being generated by every dollar of sales.

For Glitz n' Glamour this means taking the $22,500 net income from the income statement and dividing by the net sales of $860,000 from the income statement.

$$return\ on\ sales = \frac{net\ income}{net\ sales} = \frac{\$22,500}{\$860,000} = 2.61\%$$

This means for every $100 in sales, the firm had a profit of $2.61. Between 4% and 5% is considered a reasonable return; so in its return on sales, Glitz n' Glamour needs to identify some ways to increase this percentage.

Return on Assets

Return on assets is net income divided by total assets. This information helps the company understand how well they are using their assets to generate profits.

For Glitz n' glamour this equates to $22,500 in net income from the income statement, and $519,000 by adding up the company's assets for fiscal year 2018 from the balance sheet.

$$return\ on\ assets = \frac{net\ income}{total\ assets} = \frac{\$22,500}{\$519,000} = 4.33\%$$

A return of $4.33 on every $100 invested in assets is quite a low return, which suggests Glitz n' Glamour has an opportunity to improve how they utilize their assets to generate profits for the company.

Return on Owners' Equity

Return on owners' equity, also called return on investment (ROI), is net income divided by owners' equity. This information helps a company understand how much profit the company is returning to shareholders based on their invested capital.

Glitz n' Glamour generates $35.15 of profit for every $100 invested in the business.

$$return\ on\ owners'\ equity = \frac{net\ income}{owners'\ equity} = \frac{\$22,500}{\$64,000} = 35.15\%$$

Because the average for business in general is 12% to 15%, Glitz n' Glamour is delivering a good return to its owners (investors or stockholders).

Decisions

Hi, Chitra! Thanks for seeing me on such short notice.

No problem at all, Naomi. What did you want to talk about?

It seems like it's coming back to the same problem I faced before. I seem to be selling lots of merchandise, but I just don't seem to be making as much money in the end as I thought.

This sounds like you are talking about _____, this is also known as gross margin.

A. return on sales

B. return on assets

C. retrun on profit

D. return on equity

I guess? What does this really mean though?

If you divide your net income, which you think is too low, by your net sales, which you think is actually quite high, then you will arrive your return on sales. This tells you _____.

A. how much money you are making overall

B. how much profit is being generated by each dollar of sales

C. how many sales you are achieving for each dollar of profit you earn

D. how much profit is being generated by each dollar of assets you own

Okay, but why is this number important? Don't I already know how much money I'm making?

Because it starts to show you if you are actually profiting from all of your sales. If not, then you know that either _____ or _____.

A. your expenses are too low; prices are not high enough

B. your expenses are too high; prices are too high

C. your expenses are too low; prices are too high

D. your expenses are too high; prices are not high enough

Really? You know that might really help me narrow down on what I should actually do about this problem!

Yes! See, this is fairly straight forward, and these financial ratios can help paint a picture of the health of your business.

Indeed! Let's get painting, haha!

Correct Answers: A; B; D

Accounting and Financial Statements: Test

1. Delores is a financial accountant. One of her roles is to

   A. prepare accounting information and analyses for managers and decision makers.

   B. prepare and monitor department and company-wide budgets.

   C. analyze the costs of production and marketing.

   D. manage and control inventory and production costs.

   E. perform audits to ensure accuracy of the financial information.

2. Lester, Lester & Lester, a law firm, has an accountant on staff to manage its financial information. This accountant is a(n) _____ accountant.

   A. private

   B. public

   C. organization

   D. not-for-profit

   E. community

3. Dr. Badger's Pet Store is applying for a loan from the bank. The company has liabilities of $670,000 and owners' equity of $250,000. What is the debt to owners' equity ratio for Dr. Badger's Pet Store?

   A. 4.20 percent

   B. 3.11 percent

   C. 2.68 percent

   D. 0.37 percent

   E. 1.10 percent

4. Pedro's job is to keep track of all of the company's financial transactions and record them. Pedro is

   A. a bookkeeper.

   B. a financial manager.

   C. a department head.

   D. an investor.

   E. a board member.

5. On the balance sheet, insurance would be listed a(n)

   A. marketable security.

   B. fixed liability.

   C. one-time expense.

   D. prepaid expense.

   E. fixed expense.

6. Nakia has a small accounting company with three employees. She buys each person a new computer every three years. The reduction in value of Nakia's older computers would be listed on a balance sheet as

   A. asset decay.

   B. accumulated depreciation.

   C. asset decline.

   D. inventory deterioration.

   E. equipment downgrade.

7. Akito has not yet paid the rent on his office for the month. How would this debt be labeled on a balance sheet?

   A. notes payable

   B. fixed liability

   C. accounts payable

   D. long-term liability

   E. common liability

8. A mechanic took out a mortgage from a bank to buy the garage he and his employees work out of. How would this be listed on a balance sheet?

A. notes payable

B. fixed liability

C. accounts payable

D. long-term liability

E. common liability

9. Yelena just completed her income statement. She included her company's sales revenue, operating expenses, and net income. Which principal part of an income statement did Yelena forget to include?

A. profit

B. cost of goods sold

C. taxes

D. liabilities

E. payroll expenses

10. In a cash flow statement, cash flows from investing activities would include

A. cash received from sales.

B. payment of notes payable.

C. payment of mortgage payable.

D. purchase of equipment.

E. cash paid for inventory.

11. Ratio analysis can be used to analyze a firm's performance compared to

A. its previous year's performance.

B. its objectives.

C. the industry as a whole.

D. its liquidity.

E. its profitability.

12. A low inventory turnover ratio may indicate

A. higher revenue.

B. having to replenish inventory often.

C. low customer satisfaction.

D. having too much obsolete inventory.

E. efficiency in selling inventory.

13. Tree Down Landscaping recently took out a mortgage for its new vehicle storage facility. When it prepares its balance sheet, the mortgage will be classified as

A. a fixed asset.

B. owner's equity.

C. a short-term liability.

D. an intangible asset.

E. a long-term liability.

14. A(n) _____ is a specialized record book or computer program that contains summaries of all journal transactions that are accumulated into specific categories.

A. bulletin

B. ledger

C. balance sheet

D. journal

E. transaction log

15. Second Harvest Tractors sold 10,000 shares of new stock last year. How would this be classified on a cash flow statement?

A. financing activities

B. shareholder activities

C. operating activities

D. investing activities

E. fixed activities

16. Tamarack Camping Company has a common stock balance of $50,000 and retained earnings of $20,000. It has notes payable of $100,000, current notes payable of $25,000, mortgage payable of $200,000, and accounts payable of $50,000. Using this information, calculate its debt to owners' equity ratio.

A. 53.57%

B. 10.71%

C. 42.86%

D. 60%

E. 75%

17. Alabaster Enterprises has net income of $35,000, total assets of $300,000, common stock of $50,000, and retained earnings of $25,000. Calculate its return on investment.

A. 50.00%

B. 57.50%

C. 46.00%

D. 62.50%

E. 50.00%

18. Paresh called his investors to let them know that the funds received from all sales of the firm's products reached $1 million, not counting returns or allowances. Paresh was referring to his D.

A. cost of goods sold.

B. gross profit.

C. current assets.

D. gross sales.

E. net sales.

19. Zenon is an accountant for the United Way, a charitable organization. Zenon is a(n) _____ accountant.

A. private

B. public

C. organization

D. not-for-profit

E. community

20. According to the Generally Accepted Accounting Principles (GAAP), financial statements must be relevant, consistent, reliable, and

A. timely.

B. subjective.

C. comparable.

D. understandable.

E. flexible.

A

360-degree assessment
Employees are appraised not only by their managers but also by their coworkers, subordinates, and sometimes customers or clients

absolute advantage Exists when one country has a monopoly on producing a product more cheaply or efficiently than any other country can

accelerator Accelerators are programs that help entrepreneurs bring their products into the marketplace through providing support and resources designed to help the business launch and scale operations.

accessory equipment Smaller, more mobile equipment

accountability Managers must report and justify their work results to managers above them

accounting The process of measuring, classifying, analyzing, and communicating financial information

accounting equation Assets = liabilities + owners' equity

accounting process the process of collecting, recording classifying, summarizing, reporting, and analyzing financial data

accounting process Six activities that result in converting information about

individual transactions into financial statements that can then be analyzed

accounts payable Money owed to others that the firm has not yet paid

accounts receivable The total amount owed to a firm from customers who have purchased goods or services on credit

accumulated depreciation The reduction in value of assets to reflect their wearing down or obsolescence over time

acid-test ratio (Cash + marketable securities + receivables) ÷ current liabilities (also known as *quick ratio*)

acquisition Occurs when one company buys another one

Active Corps of Executives (ACE) A Small Business Association mentoring program composed of executives who are still active in the business world but have volunteered their time and talents

activity ratios Used to evaluate how well management uses a firm's assets to generate revenue (also known as *efficiency ratios*)

actual close the salesperson concludes the presentation by asking the prospect to purchase the product

actual close The salesperson concludes the presentation

by asking the prospect to purchase the product

administrative expenses Costs incurred for the general operation of the business

advertising Paid nonpersonal communication by an identified sponsor (person or organization) using various media to inform an audience about a product

advertising media The variety of communication devices for carrying a seller's message to prospective buyers

advocacy advertising Concerned with supporting a particular opinion about an issue

affirmative action Aims to achieve equality of opportunity within an organization

agency shop Workplace in which workers must pay the equivalent of union dues, although they are not required to join the union

agents Specialists who bring buyers and sellers together and help negotiate a transaction (also known as *brokers*)

agents Tend to maintain long-term relationships with the people they represent

allowances Partial refunds to customers for damaged products they choose to keep, not return

analytic transformation The process in which resources are broken down to create finished products

angel investor Individuals who invest their own money in a private company, typically a start-up

annual report A year-end report that shows a firm's financial condition and outlook for the future

antitrust law A set of laws designed to keep markets competitive by deterring big businesses from driving out small competitors

Appellate courts Appellate courts: courts that review cases appealed from lower courts, considering questions of law but not questions of fact

application (DEI) The action of putting DEI into operation

appreciation (DEI) Recognition and enjoyment of the good qualities of DEI

apprenticeship Training program in which a new employee works with an experienced employee to master a particular craft

arbitration The process in which a neutral third party listens to both parties and makes a decision that the parties have agreed will be binding on them

arbitration The process in which a neutral third party, an arbitrator, listens to both parties in a dispute and makes a decision that the parties have agreed will be binding on them

Asia-Pacific Economic Cooperation (APEC) Common market of 21 Pacific Rim countries whose purpose is to improve economic and political ties

ask Highest selling price asked

assembly line Consists of a series of steps for assembling a product, each step using the same interchangeable parts and each being performed repetitively by the same worker

assets Anything of value that is owned by a firm

at-will employment The employer is free to dismiss any employee for any reason at all—or no reason—and the employee is equally free to quit work (also known as *employment at will*)

auction site a digital retail site that lists from individuals or firms that can be purchased through an auction bidding process or directly through a "purchase now" feature.

authority The legitimacy an organization confers on managers in their power to make decisions, give orders, and utilize resources

autocratic leaders Make decisions without consulting others

automation Using machines as much as possible rather than human labor to perform production tasks

B

B corporation Legally requires that the company adhere to socially beneficial practices, such as helping communities, employees, consumers, and the environment (also known as benefit corporation)

balance of trade The value of a country's exports compared to the value of its imports over a particular period of time

balance sheet Statement of a firm's financial condition at a given time showing its assets, liabilities, and owners' equity

bankruptcy The legal means of relief for debtors unable to pay their debts

bargain hunting digital purchasing behavior that involves coupon or auction sites. Bargain hunting is often combined with browsing and may or may not lead to a purchase.

barter To trade goods or services without the exchange of money

barter The trading of goods and/or services for other goods and/or services

base pay Consists of the basic wage or salary workers are paid for doing their jobs

benchmarking A process by which a company compares its performance with that of high-performing organizations

benefit segmentation Consists of categorizing people according to the benefits, or attributes, that people seek in a product

benefits Nonwage or nonsalary forms of compensation paid for by the organization for its employees

bid Highest price a buyer (bidder) is willing to pay

bill of materials Essentially a list of materials that go into the finished product

blue-chip stocks Preferred or common stocks of big, reputable companies, which also usually pay regular dividends

bond rating measures the quality and safety of a bond, indicating the likelihood that the debt issuer will be able to meet scheduled repayments, which dictates the interest rate paid.

bonds Contracts between issuer and buyer in which the purchase price represents a loan by the buyer and for which the issuing firm pays the buyer interest

bonds Long-term IOUs issued by governments and corporations, contracts on which the issuer pays the buyer interest at regular intervals

Bonds contracts between issuer and buyer in which the purchase price represents a loan by the buyer and for which the issuing firm pays the buyer interest

Bonds long-term IOUs issued by governments and corporations, contracts on which the issuer pays the buyer interest at regular intervals.

bonds payable Long-term liabilities that represent money lent to the firm that must be paid off

bonuses Cash awards given to employees who achieve particular performance objectives

Bonuses are cash awards given to employees who achieve particular performance objectives.

book value a company subtracts its liabilities from its assets, and the resulting figure, the shareholders'

equity, is then divided by the number of shares available of the stock.

book value A company subtracts its liabilities from its assets, and the resulting figure, the shareholders' equity, is then divided by the number of shares available of the stock

bookkeeping Recording a company's financial transactions

bootstrapping a process of funding a business by using personal funds rather than seeking debt or equity investment.

bots a software application that runs automated tasks over the Internet.

bounce rate the percentage of visitors who enter a website and then quickly depart, or bounce, rather than continuing to view other pages within the same site.

brain drain The emigration of highly skilled labor to other countries in order to better their economic condition

Brain drain the emigration of highly skilled labor to other countries in order to better their economic condition

brainstorming A process wherein individuals or members of a group generate multiple ideas and alternatives

brand A unique name, symbol, or design that identifies an organization and its product or service

brand advertising Consists of presentations that promote specific brands to ultimate consumers (also known as *product advertising*)

brand awareness Consumers recognize the product

brand equity The marketing and financial value derived from the combination of factors that people associate with a certain brand name

brand insistence Consumers insist on the product; they will accept no substitutes

brand loyalty Commitment to a particular brand—the degree to which consumers are satisfied with a product and will buy it again

brand manager Person responsible for the key elements of the marketing mix—product, price, place, and promotion—for one brand or one product line (also known as *product manager*)

brand marks The parts of a brand that cannot be expressed verbally, such as graphics and symbols

brand names The parts of a brand that can be expressed verbally, such as by words, letters, or numbers

brand preference Consumers habitually buy the product if it is easily available, but will try alternatives if they can't find it

break-even analysis A way to identify how much revenue is needed to cover the total costs of developing and selling a product

break-even point The point at which sales revenues equal costs; that is, the point at which there is no profit but also no loss

brokerage firms Companies that buy and sell stocks and bonds for individuals and offer high-interest-rate combination checking and savings accounts

Brokerage firms companies that buy and sell stocks and bonds for individuals and offer high-interest-rate combination checking and savings accounts.

brokers Usually hired on a temporary basis; relationship with the buyer or the seller ends once the transaction is completed

browsing digital purchasing behavior wherein the consumer is not really looking to make a purchase.

budget A detailed financial plan showing estimated revenues and expenses for a particular future period, usually one year

bundling The practice of pricing two or more products together as a unit

business Any activity that seeks to make a profit by satisfying needs through selling goods or services and generating revenue; a for-profit organization

business cycle The periodic but irregular pattern of ups and downs in total economic production

business environment The arena of forces (economic, technological, competitive, global, and social) that encourage or discourage the development of business

business market Consists of those business individuals and organizations that want business goods and services that will help them produce or supply their own business goods and services (also known as *business-to-business market*)

business model The needs the firm will meet, the operations of the business, the company's components and functions, and its expected revenues and expenses

business plan A document that outlines a proposed firm's goals, the methods for achieving them, and the standards for measuring success

business services Services used in operations

business-to-business market Consists of those business individuals and organizations that want business goods and services that will help them produce or supply their own business goods and services (also known as *business market*)

business-to-consumer (B2C) refers to selling products and services directly to the consumer of the products or services

buying club site a digital retail site that allows consumers to buy in bulk

buzz marketing Using high-profile entertainment, social media, or news to get people to talk about their product

C

C corporation A state-chartered entity that pays taxes and is legally distinct from its owner

callable bonds Bonds in which the issuer may call them in and pay them off at a predetermined price before the maturity date

Callable bonds bonds in which the issuer may call them in and pay them off at a

predetermined price before the maturity date.

canned presentation uses a fixed, memorized selling approach to present the product

canned presentation Uses a fixed, memorized selling approach to present the product

capital budgets Used to predict purchases of long-term assets

capital expenditures Major investments in tangible or intangible assets

capital gain The return made by selling a security for a price that is higher than the price the investor paid for it

capital items Large, long-lasting equipment

capitalism An economic system in which the production and distribution of goods and services are controlled by private individuals rather than by the government (also known as *free-market economy*)

Carroll's Global Corporate Social Responsibility Pyramid Archie Carroll's guide for thinking about the day-to-day practical and moral matters that businesses encounter; the pyramid suggests that an organization's obligations in the global economy are to be a good global corporate citizen, to be ethical, to obey the law, and to be profitable

cash balance Balance in the firm's cash account at the end of the year

cash budgets Used to predict cash shortages or surpluses during the year

Cash Budgets: used to predict cash shortages or surpluses during the year.

cash flow forecasts A prediction about how money will come into and go out of a firm in the near future

cash flow statement Reports over a period of time, first, the firm's cash receipts and, second, disbursement related to the firm's (1) operating, (2) investing, and (3) financing activities, which leads to the bottom line of (4) the cash balance

cash-and-carry wholesaler A limited-function wholesaler that sells mainly to small retailers, who come to the wholesaler, pay cash for a product, and carry it out ("cash and carry")

Cash-and-carry wholesaler A limited-function wholesaler that sells mainly to small retailers, who come to the wholesaler, pay cash for a product, and carry it out ("cash and carry").

catalog marketing Consists of mailing customers catalogs, from which they may choose merchandise to be ordered via mail, telephone, or online (also known as *mail-order marketing*)

cause marketing A commercial activity in which a business forms a partnership with a charity or nonprofit to support a worthy cause, product, or service.

cause-related marketing A commercial activity in which a business forms a partnership with a charity or nonprofit to support a worthy cause, product, or

service; also known as *cause marketing*

central-planning economies Economic systems in which the government owns most businesses and regulates the amounts, types, and prices of goods and services (also known as *command economies*)

centralized authority Important decisions are made by higher-level managers

certificate of deposit Pays interest upon the certificate's maturity date

chatter Another form of consumer feedback that occurs when a consumer shares, forwards, or "retweets" a marketing message. For marketers, the level of chatter represents consumer feedback.

checking account Allows you to deposit money in a bank account and then write checks on that account

checking account allows you to deposit money in a bank account and then write checks on that account

click path a sequence of hyperlink clicks that a website visitor follows on a given site, recorded and reviewed in the order the consumer viewed each page after clicking on the hyperlink.

closed shop An employer may hire only workers for a job who are already in a union

co-branding Two noncompeting products link their brand names together for a single product

code of ethics A written set of ethical standards to

help guide an organization's actions

Cognitive diversity Cognitive diversity utilizes the different experiences and perspectives of individuals to address particular situation, challenge, opportunity, or problem.

cold-call sales prospecting technique Consists of calling on prospects with whom you have had no previous contact and to whom you do not have any kind of introduction

collateral Asset that is pledged to secure the loan

collective bargaining Consists of negotiations between management and employees in disputes over compensation, benefits, working conditions, and job security

collective bargaining The process by which labor and management representatives meet to negotiate pay, benefits, and other work terms

command economy Economic system in which the government owns most businesses and regulates the amounts, types, and prices of goods and services (also known as *central-planning economy*)

commercial bank A federal- or state-chartered profit-seeking financial institution that accepts deposits from individuals and businesses and uses part of them to make personal, residential, and business loans

commercial finance companies Organizations willing to make short-term loans to borrowers who can offer collateral

commercial paper Unsecured, short-term promissory notes over $100,000 issued by large banks and corporations

commercialization The full-scale production and marketing of the product

commodities trading Trading in raw materials and agricultural products used to produce other goods

commodity exchange A security exchange in which futures contracts are bought and sold

common market Group of nations within a geographical region that have agreed to remove trade barriers with one another (also known as *economic community* or *trading bloc*)

common stock Stockholders are able to vote on major company decisions, but they get (1) last claim on the company's dividends and (2) last claim on any remaining assets if the company goes out of business and its assets are sold

communism An economic system in which the government owns all property and everyone works for the government

Community Development Financial Institution (CDFI) A Community Development Financial Institution (CDFI) is a private financial institution who provide investing as well as personal and business lending opportunities to underserved communities within the United States.

comparative advantage Economic principle stating that a country sells to other countries those products and services it produces most cheaply or efficiently; the country buys from other countries those goods or services that it does not produce most cheaply or efficiently

compensation and benefits Laws and administration around worker's hours, pay, and benefits

competitive advantage An organization's ability to produce goods or services more effectively than its competitors

competitive advertising Promotes a product by comparing it more favorably to rival products (also known as *comparative advertising*)

competitive pricing The strategy in which price is determined in relation to rivals, factoring in other considerations such as market dominance, number of competitors, and customer loyalty

competitor People or organizations that are rivals for a company's customers or resources

compliance-based ethics code Ethical code that attempts to prevent criminal misconduct by increasing control and by punishing violators

component parts Finished or nearly finished products for making principal product

compressed workweek An employee works a full-time job in less than five days of standard 8- or 9-hour shifts

computer-aided design (CAD) Programs that are used to design products, structures, civil engineering drawings, and maps

computer-aided manufacturing (CAM) The use of computers in the manufacturing process

computer-integrated manufacturing (CIM) Systems in which computer-aided design is united with computer-aided manufacturing

concept testing Marketing research designed to solicit initial consumer reaction to new product ideas

conceptual skills The ability to think analytically, to visualize an organization as a whole, and understand how the parts work together

consideration Promising to do a desired act or refrain from doing an act you are legally entitled to do in return for something of value, such as money

consumer buying process The five steps by which consumers make decisions when considering whether to buy a product

consumer feedback different ways that customers can report their satisfaction or dissatisfaction with a firm's products.

consumer market Consists of all those individuals or households that want goods or services for their personal use

consumer price index (CPI) An index that encapsulates the monthly costs of a "market basket" of about 400 representative consumer goods and services that allow data analysts to measure the rate of inflation or deflation

consumer review a direct assessment of a product (good, service, or idea) that is expressed through social media for others to see and consider.

consumer sovereignty The idea that consumers influence the marketplace through their decisions of which products they choose to buy or not to buy

consumer-protection laws Laws concerned with protecting buyers' rights

containerization Products are packed into 20- or 40-foot-long (by about 8-foot square) containers at the point of origin and retrieved from the containers at the point of destination

contingency planning The creation of alternative hypothetical courses of action that a company can use if its original plans don't prove workable

continuity The timing of the ads, how often they appear or how heavily they are concentrated within a time period

continuous innovation Modest improvements to an existing product to distinguish it from competitors; they require little consumer behavior change

continuous processes A production process in which goods or services are turned out in a long production run on an ongoing basis over time

control process A four-step process: (1) establish standards; (2) monitor performance; (3) compare performance against standards; and (4) take corrective action, if needed

control standard The desired performance level for a given goal

controlling Monitoring performance, comparing it with goals, and taking corrective action as needed

convenience goods and services Inexpensive products that people buy frequently and with little effort

conversion rate the percentage of users who take a desired action, such as making a purchase.

convertible bonds convertible bonds are bonds that can be converted into the issuing corporation's common stock.

convertible bonds Bonds that can be converted into the issuing corporation's common stock

convertible bonds are bonds that can be converted into the issuing corporation's common stock.

cookies data files stored on websites that can generate a profile or other data about consumers.

Cooperative Cooperative: a corporation owned by its user members, who have pooled their resources for their mutual benefit

Corporate bonds issued by businesses as a source of long-term funding, consist of secured and unsecured bonds.

corporate citizenship a concern for taking actions that will benefit society as well as the organization

corporate culture The shared beliefs and values that develop within an organization and guide the behavior of its members (also known as *organizational culture*)

Corporate policy A company's stated positions on political and social issues

corporate social responsibility (CSR) A concern for taking actions that will benefit society as well as the organization

corporation A company or group of people authorized to act as a single entity (legally a person) and recognized as such in law.

cost of capital The rate of return a firm must earn to cover the cost of generating funds in the marketplace

cost of goods sold The cost of producing a firm's merchandise for sale during a certain period

cost per thousand (CPM) The cost a particular medium charges to reach 1,000 people with an ad

cost-based pricing The strategy in which the cost of producing or buying the product—plus making a profit—is the primary basis for setting price

Cost-based pricing The strategy in which the cost of producing or buying the product—plus making a profit—is the primary basis for setting price.

countertrading Bartering goods for goods (or services)

creative selling The selling process in which salespeople determine customer needs, then explain their product's benefits to try to persuade buyers to buy the product

credit An entry recording a sum received

credit union depositor-owned, nonprofit, financial cooperatives that offer a range of banking services to their members

critical path The sequence of tasks that takes the longest time to complete

cross-functional self-managed teams Groups of workers with different skills who are given the authority to manage themselves

crowdfunding The practice of funding a project or venture by raising many small amounts of money from a large number of people, typically via the Internet

Crowdfunding Crowdfunding involves funding a project or venture by raising money from a large number of people, typically through an online site.

cultural norms The ethics, values, attitudes, and behaviors that are deemed to be normal or typical in a given culture

culture The shared set of beliefs, values, knowledge, and patterns of behavior common to a group of people

culture shock The feelings of discomfort and disorientation associated with being in an unfamiliar culture

currency Government-issued coins and paper money

currency exchange rate The rate at which one country's currency can be exchanged for the currency of another country

current assets Items that can be converted into cash within one year

current liabilities Obligations in which payments are due within one year or less

current ratio Current assets ÷ current liabilities

customer loyalty program A customer loyalty program is designed to recognize and reward loyal repeat customers with rewards, coupons, discounts, or other benefits.

customer relationship management (CRM) Emphasizes finding out everything possible about customers and then using that information to satisfy and even exceed their expectations in order to build customer loyalty over the long term

customer satisfaction The concept of offering a product to please buyers by meeting their expectations

customers People or companies that pay to use an organization's goods or services

D

dashboard a central location where all social media activity can be easily monitored.

data analysis Subjected to statistical tools to determine its significance

databases Integrated collections of data stored in computer systems

day's range Highest and lowest price for the stock during the day

debit The recording or entry of debt in an account

debt to owner's equity ratio Measures of the extent to which a company uses debt, such as bank loans, to finance its operations

debt to owners' equity ratio A measure of the extent to which a company uses debt, such as bank loans, to finance its operations; total liabilities ÷ owners' equity

decentralized authority Decisions are made by middle-level and supervisory-level managers

decision A choice made from among available alternatives

decision making Process of identifying and choosing alternative courses of action

decline stage The period in which the product falls out of favor, and the organization eventually withdraws it from the marketplace

deficit An excess of spending over revenue

Deficit an excess of spending over revenue

deflation A general decline in the prices of most goods and services

Delegation The process of assigning work to subordinates.

demand Economic concept that expresses buyers' willingness and ability to purchase goods and services at different prices

Demand curve diagram that illustrates the quantity demanded of a good at various prices

demand deposit A commercial bank's or other financial institution's checking account, from which you may make withdrawals at any time

democratic political system A political system that relies on free elections and representative assemblies

demographic segmentation Consists of categorizing consumers according to statistical characteristics of a population, such as gender, age, income, education, social class, ethnicity, and so on

demographics Measurable characteristics such as gender, age, race, and family composition

demotion When an employee's current responsibilities and pay are taken away

departmentalization The dividing up of an organization into smaller units, or departments, to facilitate management

depression A particularly severe and long-lasting recession, accompanied by falling prices (deflation)

devaluation Occurs when the value of a nation's currency is lowered relative to the value of other countries' currencies

developed countries Countries with a high level of economic development and a generally high average income level among their citizens

developing countries Countries with low economic development and low average incomes

digital mall a digital retail site where a variety of sellers stock their goods.

digital marketing online marketing that can deliver content immediately to consumers through digital channels, devices, and platforms.

digital marketplace a digital retail side made up of small, independent sellers.

direct channel A producer sells directly to consumers, using mail order, telemarketing, the Internet, and TV ads

direct mail marketing Consists of mail promotions—letters, brochures, and pamphlets—sent through the postal service to customers

direct selling Face-to-face selling directly to customers in their homes or where they work

direct-action advertising Attempts to stimulate an immediate, or relatively immediate, purchase of a product through such devices as one-day sales, one-time promotions, or announcements of a special event

Direct-action advertising attempts to stimulate an immediate, or relatively immediate, purchase of a product through such devices as one-day sales, one-time promotions, or announcements of a special event.

discipline Punishing an employee, often for a poor performance appraisal, usually by suspending or demoting that employee

discontinuous innovation The product is totally new, radically changing how people live

discount brokers Execute the buy and sell orders indicated by clients but don't offer advice and tax planning

discount rate The interest rate at which the Federal Reserve makes short-term loans to member banks

discounting Assigning regular prices to products, but then resorting to frequent price-cutting strategies, such as special sales, to undercut the prices of competitors

discretionary order An order in which the customer trusts the broker's professional experience and judgment and leave it to him or her to decide the right time and price for buying or selling a security

discrimination When people are hired or promoted—or denied hiring or promotion—for reasons not relevant to the job

distribution center Provides storage of product for the short periods of time for collection and distribution elsewhere

distribution channel A system for conveying goods or services from producers to customers

distribution mix The combination of distribution channels a company uses to get its products to customers

distribution strategy An overall plan for moving products from producer to customer

distributor A person or organization (such as a dealer or retailer) that helps sell goods and services to customers

diversification Choosing securities in such a way that a loss in one investment won't have a devastating impact on your total portfolio

diversity Diversity typically refers to the similarities and differences among individuals including dimensions of personality and identity, as well as perspective and experience.

dividend and yield Annual dividend as a percentage of the price per share

dividends Part of a company's profits that are distributed to stockholders

Division of labor Different parts of a task are done by different people.

divisional structure Employees are grouped by purpose: customer groups, geographic regions, work processes, products, or industries

door-to-door selling Salespeople call directly on people at their homes or workplaces

double-entry bookkeeping The process of recording a transaction in two different accounts in order for the books to balance as a check on errors

Dow Jones Industrial Average (DJIA) Also known as "*the Dow*," a general measure of the movement of U.S. stock prices, and an index of the average of prices of the stocks of 30 large corporations

drop shipper A limited-function wholesaler who owns (has title to) the products, but does not have physical custody of them; the drop shipper takes orders and has the producer ship the product directly to the customer

dumping Occurs when a foreign company sells its products abroad for less—even less than the cost of manufacture—than the price of the domestic product

dynamically continuous innovation Marked changes to an existing product that require a moderate amount of consumer learning or behavior change

E

e-business Using the Internet to facilitate every aspect of running a business

e-cash Money held, exchanged, and represented in electronic form and transacted over the Internet

e-commerce The buying and selling of products or services over computer networks

earned media when a business or company receives recognition or acknowledgment organically

economic community A group of nations within a geographical region that have agreed to remove trade barriers with one another (also known as *common market* or *trading bloc*)

economic responsibility Seeking to be profitable as a means to create a strong economic foundation

economics The study of the production, distribution, and consumption of scarce goods and services

economies of scale The savings realized from buying materials or manufacturing products in large quantities

editing Refers to checking over to eliminate mistakes

effective To achieve results; to realize the firm's goals by making the right decisions and executing them successfully

efficient To use people, money, raw materials, and other resources wisely and cost-effectively

electronic commerce The buying and selling of products or services over computer networks

electronic funds transfer systems (EFTSs) Computerized systems that move funds from one institution to another over electronic links

email marketing a cost-effective form of digital marketing used to retain, nurture, or attracting customers

embargo A complete ban on the import or export of certain products

Emotional Intelligence the capacity to be aware of, control, and express one's emotions, and to handle interpersonal relationship with empathy

employee benefits The benefits to which employees are entitled

employee buyout A firm's employees borrow money against their own assets, such as their houses or their pension funds, to purchase the firm from its present owners; the employees then become the new owners of the firm

employee non-compete contract The employee non-compete contract is a legally binding arrangement between an employee and their employer

employment at will The employer is free to dismiss any employee for any reason at all—or no reason—and the employee is equally free to quit work (also known as *at-will employment*)

employment tests Consist of any procedure used in the employment selection decision process

empowerment Employees share management responsibilities, including decision making

endless-chain sales prospecting technique Consists of asking each sales prospect to provide the salesperson with some names of other prospects who might be interested in the product

enterprise resource planning (ERP) A computer-based system that collects and provides information about a company's entire enterprise, including identifying customer needs, receipt of orders, distribution of finished goods, and receipt of payment

enterprise zone A specific geographic area in which government tries to attract business investment by offering lower taxes and other government support

entrepreneur A person who sees a new opportunity for a product or service and who risks time and money to start a business with the goal of making a profit

entrepreneurial team A group of people with different kinds of expertise who form a team to create a new product

entrepreneurs Business owners who see a new opportunity for a product or service and start a firm

entrepreneurship The process of taking risks to try to create a new business

environmental scanning Involves looking at the wider world to identify what matters can affect the marketing program

Equal Employment Opportunity (EEO) Commission Enforces antidiscrimination and other employment-related laws

equilibrium price Determined by the point at which quantity demanded and quantity supplied intersect (also known as market price)

equity Equity refers to fair treatment in access, opportunity, and advancement for all individuals.

equity theory Focuses on employee perceptions as to how fairly they think they are being treated compared to others

Equity theory Focuses on employee perceptions as to how fairly they think they are being treated compared to others.

ethical responsibility Taking host-country and global standards into consideration when making business decisions

ethics Principles of right and wrong that influence behavior

ethics officer Company executive whose job is to integrate the organization's ethics and values initiatives, compliance activities, and business conduct practices into the company's decision-making processes

European Union (EU) The European common market, consisting of 28 trading partners in Europe

everyday low pricing (EDLP) A strategy of continuously setting prices lower than those of competitors and then not doing any other price-cutting tactics such as special sales, rebates, and cents-off coupons

exchange-traded fund (ETF) A collection of stocks that is traded on an exchange that can be traded *throughout* the trading day

excise tax Taxes based on the value of services or property other than real estate, such as airline tickets, gasoline, and firearms; beer, liquor, and cigarettes (sin taxes); and yachts, expensive cars, and fur coats (luxury taxes)

expectancy theory Proposes that people are motivated by (1) how strongly they want something, and (2) how likely they think they are to get it

expenses Costs incurred as part of a company's operating activities

exporting Producing goods domestically and selling them outside the country

expropriation Occurs when a government seizes a domestic or foreign company's assets

external recruiting What companies do in trying to attract job applicants from outside the organization

extrinsic motivator The external payoff, such as money or recognition, a person receives from others for performing a particular task

F

facility layout The physical arrangement of equipment, offices, rooms, people, and other resources within an organization for producing goods or services

facility location The process of selecting a location for company operations

factoring accounts receivable A firm sells its accounts receivable at a discount to a financial institution

factors of production The resources used to create wealth

family brands The same brand name is given to all or most of a company's products

favorable balance of trade Exists when the value of a country's total exports exceeds the value of its total imports

fear-appeal advertising Attempts to stimulate the purchase of a product by

motivating consumers through fear of loss or harm

federal budget deficit Occurs when the federal government spends more than it collects in tax revenues

Federal Deposit Insurance Corporation (FDIC) An independent agency of the U.S. government that insures bank deposits up to $250,000

Federal Reserve System Called *the Fed*; the central bank of the United States and controls the U.S. money supply

Finance the business function of obtaining funds for a company and managing them to accomplish the company's objectives

finance companies Nondeposit companies that make short-term loans at higher interest rates to individuals or businesses that don't meet the credit requirements of regular banks

financial accounting Preparing accounting information and analyses primarily for people outside of the organization

financial budgets Concentrate on the company's financial goals and the resources needed to achieve them

financial control Process by which a company periodically compares its actual revenues and expenses with those predicted in its budget

financial leverage the technique of using borrowed funds to increase a firm's rate of return

financial leverage The technique of using borrowed funds to increase a firm's rate of return

financial management The job of acquiring funds for a firm and managing them to accomplish the firm's objectives

financial managers The people responsible for planning and controlling the acquisition and uses of funds

financial plan A document that lays out a firm's strategy for reaching its financial goals

Financial plan a firm's strategy for reaching its financial goals

financial statements A summary of all transactions occurring during a particular time period; there are three types of financial statements: balance sheets, income statements, and statements of cash flows

fiscal policy The U.S. government's attempts to stabilize the economy by (1) raising or lowering taxes, or (2) borrowing and spending money

fiscal year The 12-month period designated by a company for annual financial reporting purposes

fixed assets Items that are held for a long time and are relatively permanent

fixed costs Those expenses that don't change, no matter how many products are sold; examples might include rent, insurance, utilities, and property taxes

fixed-position layout Materials, equipment, and labor are transported to one location

flexible manufacturing system (FMS) A facility that can be modified quickly to manufacture different products

flexible time Consists of flexible working hours, or any schedule that gives an employee some choices in working hours (also known as *flextime*)

Flextime Consists of flexible working hours, or any schedule that gives an employee some choices in working hours (also known as *flexible time*)

focus group Small group of people who meet with a discussion leader and give their opinions about a product or other matters

for-profit organization An organization formed to make money, or profits, by selling goods and services

forced ranking performance review systems Systems in which all employees within a business unit are ranked against one another, and grades are distributed along some sort of bell curve; top grade earners are then rewarded with bonuses and promotions, and low grade earners are warned or dismissed

forecasting Predicting revenues, costs, and expenses for a certain period of time

Foreign Corrupt Practices Act U.S. law that makes it illegal for employees of U.S. companies to make "questionable" or "dubious" contributions to political decision makers in foreign nations

foreign licensing A company gives a foreign company permission, in return for a fee, to make or distribute the licensing company's product or service

foreign subsidiary A company in a foreign country that is totally owned and controlled by the parent company

form utility The value that people add in converting resources—natural resources, capital, human resources, entrepreneurship, and knowledge—into finished products

formal appraisals Appraisals that are conducted at scheduled times of the year and are based on pre-established performance measures

franchise An arrangement in which a business owner allows others the right to use its name and sell its goods or services within a specific geographical area

franchisee The buyer of the franchise

franchising A company allows a foreign company to pay it a fee and a share of the profit in return for using a brand name and a package of materials and services

franchisor The business owner that gives others the rights to sell its products or services

free trade The movement of goods and services among nations without political or economic restrictions

free-market economy An economic system in which the production and distribution of goods and services are controlled by private individuals rather than by the government (also known as *capitalism*)

free-rein leaders Set objectives, and employees are relatively free to choose how to achieve them.

Freelancer A freelancer is an individual who is self-employed, and utilizes their knowledge, skills, and abilities, to perform a tasks, services or functions for a customer.

frequency The average number of times each member of the audience is exposed to an ad

front-line managers Make daily operating decisions, directing the daily tasks of individual contributors and nonmanagerial personnel

full-service brokers Offer a wide range of investment-related services, not only execution of trades but also investment research, advice, and tax planning (also known as *traditional brokers*)

full-service merchant wholesalers An independently owned firm that takes title to—that is, becomes owner of—the manufacturer's products and performs all sales and distribution, as well as provides credit and other services

functional structure People performing similar activities or occupational specialties are put together in formal groups

futures contract Making an agreement with a seller or broker to buy a specific amount of a commodity at a certain price on a certain date

G

gainsharing The distribution of savings or "gains" to groups of employees that reduced costs and increased measurable productivity

Gantt chart A kind of time schedule—a specialized bar chart that shows the relationship between the kind of work tasks planned and their scheduled completion dates

Gantt Chart A kind of time schedule—a specialized bar chart that shows the relationship between the kind of work tasks planned and their scheduled completion dates.

general obligation bonds Used by tax-levying government agencies to pay for public projects that will not generate revenue, such as road repairs

general partnership Two or more partners are responsible for the business, and they share profits, liabilities (debt), and management responsibilities

Generally Accepted Accounting Principles (GAAP) A set of accounting standards used in the preparation of financial statements to ensure that they are relevant, reliable, consistent, and comparable

geographic segmentation Categorizes customers according to geographic location

geotargeting Geotargeting allows marketers to specify the location where specific ads and content will be seen

by a customer based on geographic location.

geotracking use of a consumer's geographic location to determine what goods will come up in a search and at what price.

global climate change an increase in the average temperature of Earth's atmosphere.

Global Compact A voluntary agreement established in 2000 by the United Nations that promotes human rights, good labor practices, environmental protection, and anticorruption standards for businesses

global economy The increasing interaction of the world's economies as a single market instead of many national markets

global outsourcing Using suppliers outside the United States to provide labor, goods, or services (also known as *offshoring*)

Global warming an increase in the average temperature of Earth's atmosphere.

globalization Refers to the movement of the world economy toward becoming a more interdependent system

globalization The increasing connectivity and interdependence of the world's economies, societies, and cultures because of advances in communication, technology, trade, international investment, currency movement, and migration

goal A broad, long-range target that an organization wishes to attain

goal-setting theory Proposes that employees can be motivated by goals that are specific and challenging but achievable

going public Occurs when a privately owned company becomes a publicly owned company by issuing stock for sale to the public

good A tangible product (one that you can touch)

goodwill An amount paid for a business beyond the value of its other assets, based on its reputation, customer list, loyal employees, and similar intangibles

Government bonds bonds sold by the U.S. Treasury, consist of treasury notes and treasury bonds

government regulators Government agencies that establish rules and regulations under which organizations must operate

grievance is a complaint by an employee that management has violated the terms of the labor-management agreement.

grievance is a complaint by an employee that management has violated the terms of the labor-management agreement.

grievance A complaint by an employee that management has violated the terms of the labor-management agreement

gross domestic product (GDP) The total value of all the goods and services that a country produces within its borders in one year

gross domestic product (GDP) The total value of all the goods and services that a country produces within its borders in one year

gross profit The amount remaining after the cost of goods sold is subtracted from the net sales

gross sales The funds received from all sales of the firm's products

growth stage The most profitable stage, this is the period in which customer demand increases, the product's sales grow, and later competitors may enter the market

growth stocks stocks issued by small, innovative new companies in hot industries

growth stocks Stocks issued by small, innovative new companies in hot industries

guerrilla marketing Consists of innovative, low-cost marketing schemes that try to get customers' attention in unusual ways

H

Hawthorne Effect The name given to a Harvard research group's conclusion that employees work harder if they receive added attention—if employees think managers care about their welfare and that supervisors pay special attention to them

Hero In the business sense, a person whose accomplishments embody the values of the organization.

Hierarchy of Authority An arrangement for making sure that work specialization produces the right result—that the right people do the right things at the right time.

host country The country in which a company is doing business

Hostile Environment Harassment Offensive or Intimidating Workplace.

hostile takeover Situation in which an outsider (a *corporate raider*) buys enough shares in a company to be able to take control of it against the will of the corporation's top management and directors

house-party selling A host has friends and acquaintances in for a "party" with refreshments, in return for a gift from a sponsor, who typically gives a sales presentation

human capital The productive potential of employee experience, knowledge, and actions

Human capital the productive potential of employee experience, knowledge, and actions.

Human relations movement Proposed that better human relations could increase worker productivity.

human resource (HR) management Consists of the activities managers perform to obtain and maintain an effective workforce to assist organizations in achieving goals

human skills The ability to work well in cooperation with other people to get things done

hybrid structure One in which an organization uses functional and divisional structures in different parts of the same organization

hybrid work environment A hybrid work environment typically involves an employee working a certain percentage of time in a physical office space and a certain amount of time working off-site.

I

idea generation Coming up with new product ideas, ideally by collecting ideas from as many sources as possible

ideate Ideate is used in business to reference the process of forming an idea

import quota A trade barrier that limits the quantity of a product that can be imported

importing Buying goods outside the country and reselling them domestically

inbound marketing a form of digital marketing that utilizes such tools as blogs, webinars, or follow-up emails to entice customers to a product or service offer without forcing an interaction or a purchase.

incentive A commission, bonus, profit-sharing plan, or stock option that induces employees to be more productive and attract and retain top performers

inclusion Inclusion describes the extent to which each person feels welcomed, respected, supported, and valued by a group, team, or company.

income statement Once known as the *profit-and-loss statement*, this financial statement shows a firm's revenues and expenses for a particular time period and the resulting profit or loss

income tax Taxes paid on earnings received by individuals and businesses

Income tax taxes paid on earnings received by individuals and businesses

Income tax taxes paid on earnings received by individuals and businesses

incubator A facility that offers small businesses low-cost offices with basic services

indenture terms Terms of the lending agreement

individual brands Different brand names are given to different company products

Individual Contributor Some employees are considered individual contributors. They do not have any people management responsibilities but may oversee or be accountable for managing projects, programs, or processes.

industrial goods Products used to produce other products

inflation A general increase in the prices of most goods and services

inflation A general increase in the cost of most goods and services as a result of increased prices

infomercials Extended TV commercials ranging from 2 (short form) to 28.5 (long form) minutes that are devoted exclusively to promoting a product in considerable detail

informal appraisals Appraisals that are conducted at unscheduled times and consist of less rigorous indications of employee performance

informational advertising Provides consumers with straightforward knowledge about the features of the product offered, such as basic components and price

infrastructure The set of physical facilities (including telecommunications, roads, and airports) that form the basis for a country's level of economic development

initial public offering (IPO) The first time a corporation's stock is offered for sale

innovation A product that customers perceive as being newer or better than existing products

installations Large capital purchases

institutional advertising Consists of presentations that promote a favorable image for an organization

institutional investors Large and powerful organizations such as pension funds and insurance companies, which invest their own or others' funds

insurance companies Nondeposit companies that accept payments from policyholders

Insurance companies nondeposit companies that accept payments from policyholder.

intangible assets Assets that are not physical objects but are nonetheless valuable, such as patents, trademarks, and goodwill.

integrated marketing communication (IMC) Combines all four promotional tools to execute a comprehensive, unified promotional strategy

integrity-based ethics code Ethical code that seeks to foster responsible employee conduct by creating an environment that supports ethically desirable behavior

intentional tort A willful act resulting in injury

interest groups Groups whose members try to influence businesses and governments on specific issues

interest rate The price paid for the use of money over a certain period of time

intermediaries The people or firms that move products between producer and customers

intermittent processes A production process in which finished goods or services are turned out in a series of short production runs and the machines are changed frequently to make different products

intermodal shipping Shipping that combines use of several different modes of transportation

internal recruiting What companies do when they make employees already working for the organization aware of job openings

International Monetary Fund (IMF) International organization designed to assist in smoothing the flow of money among nations; operates as a last-resort lender that makes short-term loans to countries suffering from an unfavorable balance of payments

intrapreneur Someone who works inside an existing organization who sees an opportunity for a product or service and mobilizes the organization's resources to turn the opportunity into a reality

intrinsic motivator The internal satisfaction, such as a feeling of accomplishment, a person receives from performing the particular task itself

introduction stage The stage in the product life cycle in which a new product is introduced into the marketplace

inventory The name given to goods kept in stock to be used for the production process or for sales to customers

inventory The name given to goods kept in stock to be used for the production process or for sales to customers

inventory control The system for determining the right quantity of resources and keeping track of their location and use

Inventory Control The system for determining the right quantity of resources and keeping track of their location and use

inventory turnover ratio Cost of goods sold in one year ÷ average value of inventory

investment bankers Companies that engage in buying and reselling new securities

investment-grade bonds Bonds that are relatively safe, with a low probability

of default; they have a bond rating of BBB or above

Investment-grade bonds bonds that are relatively safe, with a low probability of default; they have a bond rating of BBB or above.

invisible hand Adam Smith's term for the market forces that convert individuals' drive for prosperity into the goods and services that provide economic and social benefits to all

Invisible hand Adam Smith's term for the market forces that convert individuals' drive for prosperity into the goods and services that provide economic and social benefits to all

ISO 9000 series Quality-assurance procedures companies must install—in purchasing, manufacturing, inventory, shipping, and other areas—that can be audited by "registrars," or independent quality-assurance experts

J

job analysis Determine the basic elements of a job, using observation and analysis

job description Outlines what the holders of the job do and how and why they do it

Job enlargement Consists of increasing the number of tasks in a job to improve employee satisfaction, motivation, and quality of production.

Job enrichment Consists of creating a job with motivating factors such as recognition, responsibility, achievement, stimulating work, and advancement.

job postings Putting information about job vacancies on company websites, break-room bulletin boards, and newsletters

job rotation Consists of rotating employees through different assignments in different departments to give them a broader picture of the organization

job routine What is required in the job for which a person was hired, how the work will be evaluated, and who the immediate coworkers and managers are

job sharing Two people divide one full-time job

job sharing two people divide one full-time job. Working at Working at home with telecommunicat ions between office and home is called

Job simplification Reducing the number of tasks a worker performs.

job specification Describes the minimum qualifications people must have to perform the job successfully

joint venture A U.S. firm shares the risk and rewards of starting a new enterprise with a foreign company in a foreign country (also known as *strategic alliance*)

journal A record book or part of a computer program containing the daily record of the firm's transactions

judiciary The branch of government that oversees the court system

just-in-time (JIT) inventory control Only minimal supplies are kept on the organization's premises and others are delivered by the suppliers on an as-needed basis

K

knockoff brands Illegal imitations of brand-name products

knowledge workers People who work primarily with information or who develop and use knowledge in the workplace

L

labor unions Organizations of employees formed to protect and advance their members' interests by bargaining with management over job-related issues

laws Rules of conduct or action formally recognized as binding or enforced by a controlling authority

leading Motivating, directing, and otherwise influencing people to work hard to achieve the organization's goals

lean manufacturing The production of products by eliminating unnecessary steps and using the fewest resources, while continually striving for improvement

ledger A specialized record book or computer program that contains summaries of all journal transactions classified into specific categories

legal responsibility Obeying the laws of host countries as well as international law

less-developed countries Countries with low economic development and low average incomes

leveraged buyout (LBO) Occurs when one firm borrows money to buy another firm; the purchaser uses the assets of the company being acquired as security for the loan being used to finance the purchase

liabilities Debts owed by a firm to an outside individual or organization

limit order Telling a broker to buy a particular security only if it is less than a certain price or to sell it only if it is above a certain price

limited liability company (LLC) Combines the tax benefits of a sole proprietorship or partnership—one level of tax—with the limited liability of a corporation

limited liability partnership (LLP) Each partner's liability—and risk of losing personal assets—is limited to just his or her own acts and omissions and those of his or her directly reporting employees

limited partnership One or more general partners plus other, limited partners who contribute an investment but do not have any management responsibility or liability

limited-function merchant wholesaler An independently owned firm that takes title to—becomes owner of—the manufacturer's products but performs only selected services, such as storage only

line managers Involved directly in an organization's goals, have authority to make decisions, and usually have people reporting to them

line of credit How much a bank is willing to lend the borrower during a specified period of time

liquidity The essential feature of current assets, as they are easily converted into cash

liquidity ratios Measure a firm's ability to meet its short-term obligations when they become due

load funds Commission is charged for each purchase

Load funds a commission is charged for each purchase.

logistics Planning and implementing the details of moving raw materials, finished goods, and related information along the supply chain, from origin to points of consumption to meet customer requirements

long-term forecasts Predictions for the next 1, 5, or 10 years

long-term liabilities obligations in which payments are due in one year or more

long-term liabilities Obligations in which payments are due in one year or more, such as for a long-term loan from a bank or insurance company

loss Occurs when business expenses exceed revenues

loss leaders Products priced at or below cost to attract customers

M

M1 The narrowest definition of the money supply, money that can be accessed quickly and easily

M1 the narrowest definition of the money supply, is defined as money that can be accessed quickly and easily

M2 (1) Money that can be accessed quickly and easily

(that is, M1) *and* (2) money that takes more time to access

M2 (1) money that can be accessed quickly and easily (that is, M1) AND (2) money that takes more time to access

macroeconomics The study of large economic units, such as the operations of a nation's economy and the effect on it of government policies and allocation of resources

management Planning, leading, organizing, and controlling the activities of an enterprise according to certain policies to achieve certain objectives

management The pursuit of organizational goals effectively and efficiently through (1) planning, (2) organization, (3) leading, and (4) controlling the organization's resources

Management by Objectives (MBO) A four-stage process in which a manager and employee jointly set objectives for the employee, manager develops an action plan for achieving the objective, manager and employee periodically review the employee's performance, and manager makes a performance appraisal and rewards employee according to results

managerial accounting Preparing accounting information and analyses for managers and other decision makers inside an organization

manufacturer-owned wholesaler A wholesale business that is owned and operated by a product's manufacturer

Manufacturer's branch office An office that is owned and managed by a manufacturer that not only has offices for sales representatives, but also carries an inventory from which the staff can fill orders.

manufacturer's brands An individual company creates a product or service and brands the product or service using the company name; also called *national* or *producer brands*, or even *global brands* when extended worldwide

Manufacturer's sales office An office that is owned and managed by a manufacturer and that has offices for sales representatives who sell products that are delivered at a later time.

Market Opportunity A market opportunity is an analysis of the potential need or want of a particular product or service.

market order Telling a broker to buy or sell a particular security at the best available price

market price Determined by the point at which quantity demanded and quantity supplied intersect (also known as *equilibrium price*)

market segmentation Divides a market into groups whose members have similar characteristics or wants and needs

market share The percentage of the market of total sales for a particular product or good

market value market value is the price at which a stock is currently selling t

market value The price at which a stock is currently selling

marketable securities Stocks, bonds, government securities, and money market certificates, which can be easily converted to cash

marketing The activity, set of institutions, and processes for creating, communicating, delivering, and exchanging offerings that have value for customers, clients, partners, and society at large

marketing concept Focuses on customer satisfaction, service, and profitability

marketing environment Consists of the outside forces that can influence the success of marketing programs. These forces are (1) global, (2) economic, (3) sociocultural, (4) technological, (5) competitive, (6) political, and (7) legal and regulatory

marketing mix Consists of the four key strategy considerations called the 4 Ps: product, pricing, place, and promotion strategies. Specifically, the marketing mix involves (1) developing a product that will fill consumer wants, (2) pricing the product, (3) distributing the product to a place where consumers will buy it, and (4) promoting the product

marketing research The systematic gathering and analyzing of data about problems relating to the marketing of goods and services

marketing strategy A plan for (1) identifying the target market among market segments, (2) creating the right marketing mix to reach that target market, and (3) dealing with important forces in the external marketing environment

Maslow's Hierarchy of Needs theory Proposes that people are motivated by five levels of needs, ranging from low to high: (1) physiological, (2) safety, (3) social, (4) esteem, and (5) self-actualization

Mass customization Using mass production techniques to produce customized goods or services

mass production The production of uniform goods in great quantities

master budgets Used to pull together the other budgets into an overall plan of action

master limited partnership (MLP) The partnership acts like a corporation, selling stock on a stock exchange, but it is taxed like a partnership, paying a lower rate than the corporate income tax

materials handling The physical handling of goods to and from and within warehouses

materials requirement planning (MRP) A computer-based method of delivering the right amounts of supplies to the right place at the right time for the production of goods

matrix structure One which combines, in grid form, the functional chain of command and the divisional chain of command—usually product—

so that there is a vertical command structure and a horizontal command structure

maturity date the issuing organization is legally required to repay the bond's principal in full to the bondholder.

maturity stage The period in which the product starts to fall out of favor and sales and profits start to level off

mechanization The use of machines to do the work formerly performed by people

media planning The process of choosing the exact kinds of media to be used for an advertising campaign

mediation The process in which a neutral third party listens to both sides in a dispute, makes suggestions, and encourages them to agree on a solution without the need for a court trial

mediation The process in which a neutral third party, a mediator, listens to both sides in a dispute, makes suggestions, and encourages them to agree on a solution

Mediation is the process in which a neutral third party, a mediator, listens to both sides in a dispute, makes suggestions, and encourages them to agree on a solution.

medium of exchange Characteristic of money that makes economic transactions easier and eliminates the need to barter

mentor An experienced person who coaches and guides lesser-experienced people by helping them understand an industry or

organization's culture and structure

mentoring The process by which an experienced employee, the mentor, supervises, teaches, and provides guidance for a less-experienced employee, the mentee or protégé

Mentoring describe s the process in which an experienced employee, the mentor, supervises, teaches, and provides guidance for a lessexperienced employee, the mentee or protégé

Mercosur The largest common market in Latin America, with 13 member countries at different levels of participation (full, associate, observer)

merger Occurs when two firms join to form a new firm

microeconomics The study of small economic units, the operations of particular groups of people, businesses, organizations, and markets

Microloan a small sum of money lent at low interest to a new business

micropreneur A person who takes the risk of starting and managing a business that remains small (often home-based)

middle managers Implement the policies and plans of the top managers above them and supervise and coordinate the activities of the supervisory managers below them

mission statement A statement of the organization's fundamental purposes

mixed economy A blended economic system in which

some resources are allocated by the free market and some resources are allocated by the government, resulting in a somewhat better balance between freedom and economic equality

mobile marketing a set of practices that enables organizations to communicate and engage with their audience in an interactive and relevant manner through and with any mobile device or network.

mobile payment A transaction completed through a portable electronic device via an application.

modular structure One in which a firm assembles pieces, or modules, of a product provided by outside contractors

monetary policy The U.S. government's attempts to manage the money supply and interest rates in order to influence economic activity

money Any medium of value that is generally accepted as payment for goods and services

money market accounts Offer interest rates competitive with those of brokerage firms but they require higher minimum balances and limit check writing

Money market accounts offer interest rates competitive with those of brokerage but they require higher minimum balances and limit check writing.

money market instruments Short-term IOUs, debt securities that mature

within one year, which are issued by governments, large corporations, and financial institutions

Money market instruments short-term IOUs, debt securities that mature within one year, which are issued by governments, large corporations, and financial institutions.

money supply the amount of money the Federal Reserve System makes available for people to buy goods and services.

money supply The amount of money the Federal Reserve System makes available for people to buy goods and services

monopolistic competition A type of free market that has many sellers who sell similar products, but the sellers have found ways to distinguish among their products, or buyers perceive the products as being different

monopoly A type of free market in which there is only one seller and no competition

mortgage A loan in which property or real estate is used as collateral

motivation The psychological processes that induce people to pursue goals

multilevel marketing Independent businesspeople, or distributors, sell products both on their own and by recruiting, motivating, supplying, and training others to sell those products, with the distributors' compensation being based on both their personal sales and the group's sales

multinational corporations Organizations with multinational management and ownership that manufacture and market products in many different countries

Municipal bonds bonds issued by state and local governments and agencies, consist of revenue bonds and general obligation bonds.

mutual fund A fund operated by an investment company that brings together money from many people and invests it in an array of diversified stocks, bonds, or other securities

mutual savings bank For-profit financial institutions similar to savings and loans, except that they are owned by their depositors rather than by shareholders

Mutual savings bank for-profit financial institutions similar to savings and loans, except that they are owned by their depositors rather than by shareholders.

N

narrow span of control Limited number of people reporting to a manager

NASDAQ Composite Index Tracks not only domestic but also foreign common stocks traded on the NASDAQ exchange

National Credit Union Administration (NCUA) An independent agency that provides up to $250,000 insurance coverage per individual per credit union

national debt The amount of money the government owes

because federal spending has exceeded federal revenue

National Labor Relations Board Enforces procedures allowing employees to vote to have a union and the rules for collective bargaining

necessity entrepreneur An entrepreneur who suddenly must earn a living and is simply trying to replace lost income

need-satisfaction presentation consists of determining customer needs and then tailoring your presentation to address those needs

need-satisfaction presentation Consists of determining customer needs and then tailoring your presentation to address those needs

negligence An unintentional act that results in injury

net income The firm's profit or loss after paying income taxes; net income = revenue − expenses

net period Length of time for which the supplier extends credit

net sales The money resulting after sales returns and allowances are subtracted from gross sales

neuromarketing The study of how people's brains respond to advertising and other brand-related messages by scientifically monitoring brainwave activity, eye tracking, and skin response

new product A product that either (1) is a significant improvement over existing products or (2) performs a new function for the consumer

niche marketing Consists of dividing market segments even further, to microsegments for which sales may be profitable

no-load funds There is no sales charge but the investment company may charge a management fee

No-load funds there is no sales charge but the investment company may charge a management fee.

non-disclosure agreement A non-disclosure agreement binds the individuals from disclosing proprietary information owned by one of the parties.

nonbanks Financial institutions—insurance companies, pension funds, finance companies, and brokerage firms—that offer many of the same services as banks provide

nonprofit organization An organization that exists to earn enough profit to cover its expenses and further its goals

nonstore retailers sell merchandise in ways other than through retail stores

nonverbal communication Messages sent outside of the written or spoken word, such as what constitutes permissible interpersonal space

North American Free Trade Agreement (NAFTA) Agreement that established a common market consisting of the 450 million people of the United States, Canada, and Mexico

not-for-profit accountants Those who work for governments and nonprofit organizations, perform the same services as for-profit accountants—except they are concerned with efficiency, not profits

notes payable Money owed on a loan based on a promise (either short term or long term) the firm made

NOW account Pays interest and allows you to write an unlimited number of checks, but you have to maintain a minimum monthly balance

O

objective A specific, short-term target designed to achieve the organization's goals

objective appraisals Appraisals that are based on facts and often based on numbers related to employees

off-the-job training Consists of classroom programs, videotapes, workbooks, online distance learning programs, and similar training tools

oligopoly A type of free market that has a few sellers offering similar but not identical products to many small buyers

on-the-job training Takes place in the workplace while employees are working at job-related tasks

onboarding Process that is designed to help a newcomer fit smoothly into the job and the organization

one-time shopping digital purchasing behavior that may combine product-focused shopping, browsing, researching, and bargain hunting all at the same time. Consumers are shopping for a gift or using a gift card and will not return to the shop once the purchase is made.

one-to-one marketing Consists of reducing market segmentation to the smallest part—individual customers

online retailing Nonstore retailing of products directly to consumers using the Internet

Online retailing Nonstore retailing of products directly to consumers using the Internet.

open shop Workers may choose to join or not join a union

open-market operations The Federal Reserve controls the money supply by buying and selling U.S. Treasury securities, or government bonds, to the public

operating budgets Used to predict sales and production goals and the costs required to meet them

operating expenses Selling and administrative expenses

operational planning Determining how to accomplish specific tasks with existing resources within the next one-week to one-year period

operations Any process that takes basic resources and converts them into finished products—inputs into outputs (also known as *production*)

operations management The management of the process of transforming materials, labor, and other resources into goods and/or services

opportunities Favorable circumstances that present possibilities for progress beyond existing goals

opportunity entrepreneur An ambitious entrepreneur who starts a business to pursue an opportunity (and large profits)

order processing Consists of receiving customer orders and seeing that they are handled correctly and that the product is delivered

organization A group of people who work together to accomplish a specific purpose

organization's mission and operations The purpose, products, operations, and history of an organization

organizational chart A box-and-lines illustration of the formal lines of authority and the official positions or work specializations

organizational culture The shared beliefs and values that develop within an organization and guide the behavior of its members (also known as *corporate culture*)

organizing Arranging tasks, people, and other resources to accomplish the work

outsourcing Using suppliers outside the company to provide goods and services (also known as *contract manufacturing*)

owned media the marketing channels that a company develops

owners Those who can claim the organization as their legal property

owners' equity Represents the value of a firm if its assets were sold and its debts paid

(also known as *stockholders' equity*)

P

packaging The covering or wrapping around a product that protects and promotes the product

packaging Covering or wrapping that protects and promotes a product

paid display advertising Includes everything from banner ads to YouTube video advertising. These ads generate awareness as well as (hopefully) drive traffic to a website.

paid media all the online marketing channels that the business pays for

paid search online advertising in which a company pays to be a sponsored result of a customer's Web search.

paid stories ads that appear as content designed to look like stories to the viewer.

paid stories ads that appear as content designed to look like stories to the viewer.

par value the face value of a share of stock, an arbitrary figure set by the issuing corporation's board of directors agment

par value The face value of a share of stock, an arbitrary figure set by the issuing corporation's board of directors

Paris agreement The Paris Agreement is a bridge between today's policies and climate-neutrality before the end of the century.

Source: https://ec.europa.eu/
clima/policies/international/
negotiations/paris_en

part-time work Any work done on a schedule less than the standard 40-hour workweek

participation to take a meaningful and active role in organizational activities

participative leaders Delegate authority and involve employees in their decisions.

partnership A business owned and operated by two or more persons as a voluntary legal association

pay for performance bases pay on the employee's work results.

pay for performance Bases pay on the employee's work results

peak The point at which an economic expansion starts to lose steam

peer-to-peer lending The process of obtaining financing from other individuals instead of a traditional financial institution like a bank or credit union.

penetration pricing Setting a low price to attract many customers and deter competition; designed to generate customers' interest and stimulate them to try out new products

pension funds Nondeposit institutions that provide retirement benefits to workers and their families

perfect competition A type of free market that has many small sellers who sell interchangeable products to many informed buyers, and no seller is large enough to dictate the price of the product

performance appraisal Consists of a manager's assessing an employee's performance and providing feedback (also known as *performance review*)

performance review Consists of a manager's assessing an employee's performance and providing feedback (also known as *performance appraisal*)

personal selling In-person, face-to-face communication and promotion to influence customers to buy goods and services

personal selling Face-to-face communication and promotion to influence customers to buy goods and services

persuasive advertising Tries to develop a desire among consumers for the product

PERT chart A diagram for determining the best sequencing of tasks

philanthropic responsibilities Acting as a good global corporate citizen, contributing resources to the community, and seeking to improve the quality of life for individuals as defined by the host country's expectations

philanthropy Charitable donations to benefit humankind

physical distribution All of the activities required to move products from the manufacturer to the final buyer

piece rate Employees are paid according to how much output they produce

Pitch Deck A summary of the company, future vision, market opportunity and business plan delivered via a presentation software consisting of typically between 10–20 slides.

placing The process of moving goods or services from the seller to prospective buyers (also known as *distribution*)

planning Setting goals and deciding how to achieve them

pledging accounts receivable A firm uses its accounts receivable as collateral, or security, to obtain a short-term loan

portfolio The collection of securities representing a person's investments

preferred stock Stockholders are not able to vote on major company decisions, but they get (1) preferred, or first, claim on the company's dividends and (2) first claim on any remaining assets if the firm goes bankrupt and its assets are sold

prepaid expense An expenditure paid for during an accounting period, but the asset will not be consumed until a later time

press release A brief statement written in the form of a news story or a video program that is released to the mass media to try to get favorable publicity for a firm or its products (also known as a *news release* or *publicity release*)

price skimming Setting a high price to make a large

profit; it can work when there is little competition

price/earnings ratio Price of stock divided by firm's per-share earnings

pricing Figuring out how much to charge for a product

pricing objectives Goals that product producers—as well as retailers and wholesalers—hope to achieve in pricing products for sale

primary data Data derived from original research, such as that which you might conduct yourself

primary securities market The financial market in which new security issues are first sold to investors

principle of motion economy Every job can be broken down into a series of elementary motions

Principle of motion economy Every job can be broken down into a series of elementary motions.

Principle of motion economy Every job can be broken down into a series of elementary motions.

private accountants In-house accountants that work for a single organization; sometimes called corporate accountants

private placements Selling stock to only a small group of large investors

private-label brands Brands attached to products distributed by one store or a chain

problems Difficulties that impede the achievement of goals

process layout Similar work is grouped by function

process materials Materials for making principal product that are not readily identifiable in that product

producer price index (PPI) A measure of prices at the wholesale level (wholesale goods are those purchased in large quantities for resale)

product A good (tangible) or service (intangible) that can satisfy customer needs

product analysis Doing cost estimates to calculate the product's possible profitability

product development The production of a prototype of the product, a preliminary version, so the company can see what the product will look like

product differentiation The attempt to design a product in a way that will make it be perceived differently enough from competitors' products that it will attract consumers

product layout Equipment and tasks are arranged into an assembly line—a sequence of steps for producing a single product in one location

Product Layout Equipment and tasks are arranged into an assembly line—a sequence of steps for producing a single product to one location.

product life cycle A model that graphs the four stages that a product or service goes through during the "life" of its marketability: (1) introduction, (2) growth, (3) maturity, (4) decline

product line A collection of products designed for a similar market, or a collection of products that are physically similar

product mix The combination of all product lines that a company offers

product placement in which sellers of a product pay to have that product prominently placed in a TV show or film so that many people will see it.[iii]

[iii] E. Porter, "Would You Buy a New Car from Eva Longoria?" *New York Times*, July 10, 2008, p. A22; S. Clifford, "Product Placements Acquire a Life of Their Own on Shows," *New York Times*, July 14, 2008, pp. C1, C4; and S. Clifford, "A Product's Place Is on the Set," *New York Times*, July 22, 2008, pp. C1, C6.

product screening Elimination of product ideas that are not feasible

product-focused shopping digital purchasing behavior that involves replacing an existing product or purchasing a product that has been pre-chosen.

production Any process that takes basic resources and converts them into finished products—inputs into outputs (also known as *operations*)

production management The management of the process of transforming materials, labor, and other resources into goods

productivity The amount of output produced for each unit of input

Productivity The amount of output per input

profit Revenue minus expenses; the amount of money a business makes after paying for all its costs

profit sharing Sharing a percentage of the company's profits with employees

profitability ratios Used to measure how well profits are doing in relation to the firm's sales, assets, or owners' equity

promissory note A written contract prepared by the buyer who agrees to pay the seller a certain amount by a certain time

promotion Moving the employee to a higher management job within the company

promotion Consists of all the techniques companies use to motivate consumers to buy their products

promotion mix The combination of tools that a company uses to promote a product, selecting from among four promotional tools: (1) advertising, (2) public relations, (3) personal selling, and (4) sales promotion

property Anything of value for which a person or firm has right of ownership

property tax Taxes paid on real estate owned by individuals and businesses, as well as on certain kinds of personal property

proportional relationship Two quantities are in a proportional relationship if they have a constant ratio, or if the graph of the quantities on a coordinate plane is a straight line through the origin.

prospecting The process of identifying potential customers, who are called *prospects*

prototype A preliminary version of a product

psychographic segmentation Consists of categorizing people according to lifestyle, values, and psychological characteristics

psychological pricing The technique of pricing products or services in odd, rather than even, amounts to make products seem less expensive

Psychological safety An environment where people believe candor is welcome

public accountants Professionals who provide accounting services to clients on a fee basis

public offerings Selling stock to the general public in securities market

public relations (PR) Unpaid, nonpersonal communication that is concerned with creating and maintaining a favorable image of the firm, its products, and its actions with the mass media, consumers, and the public at large

public service advertising Consists of presentations, usually sponsored by nonprofit organizations, that are concerned with the welfare of the community in general; such ads are often presented by the media free of charge

publicity Unpaid coverage by the mass media about a firm or its products

publicly traded company A publicly-traded company is a company that issues stock that is traded on the open market typically through a stock exchange.

pull promotional strategy Aimed directly at consumers, to get them to demand the product from retailers

purchasing The activity of finding the best resources for the best price from the best suppliers to produce the best goods and services

push promotional strategy is aimed at wholesalers and retailers, to encourage them to market the product to consumers.

push promotional strategy Aimed at wholesalers and retailers, to encourage them to market the product to consumers

Q

quadratic equation an equation containing a single variable of degree 2. Its general form is $ax^2 + bx + c = 0$, where x is the variable and $a, b,$ and c are constants ($a \neq 0$).

qualifying Determining if the prospect has the authority to buy and the ability to pay

quality Refers to the total ability of a product or service to meet customer needs

quality assurance The process of minimizing errors by managing each stage of production

quality of life The level of a society's general well-being as measured by several key factors, including health care, educational opportunities, and environmental health; also called *human development*

R

rack jobber A limited-function wholesaler who furnishes products and display racks or shelves in retail stores and shares profits with retailers

ratio analysis Uses one of a number of financial ratios—such as liquidity, efficiency, leverage, and profitability—to evaluate variables in a financial statement

raw materials Basic materials for making principal product

reach The number of people within a given population that an ad will reach at least once

recession Two or more consecutive quarters of decline in gross domestic product (GDP)

recruiting The process by which companies find and attract qualified applicants for open jobs

referral sales prospecting technique Consists of asking satisfied customers to provide names of potential customers and to contact them on behalf of the salesperson

reinforcement theory Suggests that behavior with positive consequences tends to be repeated, whereas behavior with negative consequences tends not to be repeated

reliability Expresses how well a test measures the same thing consistently

reminder advertising Tries to remind consumers of the existence of a product

remote work environment A work environment in which employees do not work at a physical site such as an office, but work from an alternative space often a home office.

representation The action of speaking or acting on behalf

of someone, or the state of being represented.

researching digital purchasing behavior wherein the consumer is purchasing a product for the first time. Unlike browsing, which has no expected outcome, research is more deliberate and will likely result in a purchase either online or offline.

reserve requirement The percentage of total checking and savings deposits that a bank must keep as cash in its vault or in a non-interest-bearing deposit at its regional Federal Reserve bank

resource development The study of how to develop the resources for creating and best utilizing goods and services

responsibility The obligation to perform the tasks assigned to you

retailers Intermediaries who sell products to the final customer

retained earnings The portion of the company profits that the owners choose to reinvest in the company

return on assets Net income ÷ total assets; this information helps the company understand how well they are using their assets to generate profits

return on owners' equity Net income ÷ owners' equity (also known as *return on investment (ROI)*)

return on sales Net income ÷ sales (also known as *profit margin*)

revenue The total amount of money that the selling of

goods or services produces during a defined period of time (for example, one year)

revenue bonds Used to pay for public projects that will generate revenue, such as toll bridges

revolving credit agreement The bank guarantees the loan and is obligated to loan funds up to the credit limit

Revolving credit agreement the bank guarantees the loan and is obligated to

risk The possibility that the owner(s) of a business may invest time and money in an enterprise that fails (that is, does not make a profit)

risk-return trade-off Financial managers continually try to balance the firm's investment risk with the expected return from its investments

Rites and rituals The activities and ceremonies, planned and unplanned, that celebrate important occasions and accomplishments in the organization's life.

robotics The use of programmable machines

robots Programmable machines used to manipulate materials and tools to perform a variety of tasks

S

S corporation Has no more than 100 owners (shareholders), but, like a partnership, the owners are taxed only at the personal level, not the corporate level

sales commission Salespeople are paid a percentage of the earnings the company made from their sales

sales promotion Short-term marketing incentives to stimulate dealer interest and consumer buying

sales returns Products that customers return to the company for a refund

sales revenue The funds received from the sales of goods and services during a certain period

sales support Consists not of selling products but of facilitating the sale by providing supportive services

Sales support consists not of selling products but of facilitating the sale by providing supportive services

sales tax Taxes collected by retail merchants on merchandise they sell

Sarbanes–Oxley Act U.S. law, enacted in 2002, that established protections for whistleblowers, recordkeeping requirements for public companies, and penalties for noncompliance

savings account A bank account that pays low interest and doesn't allow check writing

savings and loan associations (S&Ls) Financial institutions that accept deposits and were originally intended to make loans primarily for home mortgages

Savings Association Insurance Fund (SAIF) Insures depositors with accounts in savings and loan associations up to $250,000 per depositor per bank

scheduling The act of determining time periods for each task in the production process

Scientific Management Emphasized the scientific study of work methods to improve the productivity of individual workers.

Score A Small Business Association mentoring program consisting of retired executives who volunteer as consultants to advise small—business people

SCORE A Small Business Association mentoring program consisting of retired executives who volunteer as consultants to advise small business people

search engine optimization (SEO) the process of driving traffic to a company's website from "free" or "organic" search results using search engines.

secondary data Information acquired and published by others

secondary securities market The financial market in which existing stocks and bonds are bought and sold by investors

secured bonds Backed by pledges of assets (collateral) to the bondholders

secured loan The borrower pledges some sort of asset, such as personal property, that is forfeited if the loan is not repaid

securities Financial instruments such as stocks and bonds

selection process Screens job applicants to hire the best candidate

self-assessment Employees rank their own performance to become involved in the evaluation process and to make them more receptive to feedback

self-sufficiency A country's ability to produce all of the products and services it needs or that its people want; no country is self-sufficient

selling The exchange of goods or services for an agreed sum of money

selling expenses Expenses incurred in marketing the firm's products

Selling Expenses Selling Expenses are all the expenses incurred in marketing the firm's products, such as salespeople's salaries, advertising, and supplies.

sentiment analysis a measurement that indicates whether people are reacting favorably or unfavorably to products or marketing efforts.

serial bonds Bonds that mature at different dates

Serial bonds bonds that mature at different dates.

Serial bonds bonds that mature at different dates.

Servant leader leadership philosophy and set of practices in which a leader strives to serve others by enriching the lives of individuals building better organizations, and ultimately creating a more just and caring world.

service An intangible product; usually a task that is performed for the purchaser

Service Corps of Retired Executives (SCORE) A Small Business Association mentoring program consisting of retired executives who volunteer as consultants to advise small businesspeople

Service Level agreements Service Level agreements often called SLAs are contracts with customers that identify what the customer will receive, when the customer will receive it, the level of quality, and the cost.

sexual harassment Consists of unwanted sexual attention that creates an adverse work environment

shadowing An employee being trained on the job learns skills by watching more experienced employees perform their jobs

shareholders Those who own stock in a company

shopping goods and services Expensive products that people buy after comparing for value, price, quality, and style

short-term forecasts Predictions for the next year or less

Side Hustle A side hustle is a project or business that an individual works on outside of their regular employment.

sinking-fund bonds Bonds in which the issuer makes annual deposits to a bank to accumulate funds for paying off the bonds on maturity

Sinking-fund bonds bonds in which the issuer makes annual deposits to a bank to accumulate funds for paying off the bonds on maturity.

Six Sigma A rigorous statistical analysis process that reduces defects in manufacturing and service-related processes

Skunkworks Skunkworks: a team whose members are separated from an organization's normal operation and asked to produce a new, innovative project

small business In the United States, a business that (a) is independently owned and operated, (b) is not dominant in its field of operation, and meets certain criteria set by the Small Business Administration for (c) number of employees and (d) annual sales revenue

Small Business Administration (SBA) The principal U.S. government agency charged with aiding small businesses by providing help in financing, management training, and support in securing government contracts

social audit A systematic assessment of a company's performance in implementing socially responsible programs, often based on predefined goals

social entrepreneurship An innovative, social value–creating activity that can occur within or across the for-profit and nonprofit sectors

social media influencers consumers who have a large following and credibility within a certain market segment.

social media marketing one of the most popular forms of digital marketing that utilizes online social networks and applications as a method to communicate mass and personalized messages about brands and products.

social media marketing campaign a coordinated marketing effort to advance marketing goals using one or more social media platforms.

social media platform a website-based media channel used to facilitate communication and connection.

socialism An economic system in which the government owns some major industries, but individuals own smaller businesses; the government redistributes much of the wealth or surplus of high incomes through social programs

sole proprietorship A business owned, and typically managed, by one person

solopreneur Business owners who work and operate their business alone

solvency Being able to pay debts when they become due

Span of control The number of people reporting to a particular manager.

special-interest group A group whose members try to influence businesses and governments on specific issues

specialty goods and services Very expensive products that buyers seldom purchase or that have unique characteristics that require people to make a special effort to obtain them

speculative-grade bonds High-risk bonds with a greater probability of default

Speculative-grade bonds high-risk bonds with a greater probability of default

sponsorship Firms that often sponsor YouTube or Instagram celebrities who in turn endorse the firms' products. These so-called online influencers are often compensated in multiple ways for their endorsements.

staff personnel Have advisory duties; they provide advice, recommendations, and research to line managers

staffing The recruitment, hiring, motivating, and retention of valuable employees

stakeholders Those who have any sort of stake or interest in a business

Standard & Poor's 500 (S&P 500) An index of stock prices for 500 major corporations in a range of industries

standard of living A component of a society's quality of life, defined by how many goods and services people can buy with the money they have

standard of value It can be used as a common standard to measure the values of goods and services

standardization The use of uniform parts that could be easily interchanged with similar parts

statistical process control A statistical technique that uses periodic random samples from production runs to see if quality is being maintained within a standard range of acceptability

stock Shares of ownership in a company

stock certificate A paper certificate listing the shareholder's name, name of the issuing company, number of shares you hold, and type of stock being issued

stock market indicators Indexes of stock market prices of groups of stocks that are related in some way

stock options Key employees are given the right to buy stock at a future date for a discounted price

stock split a company divides its existing shares into multiple shares.

stock split A company divides its existing shares into multiple shares

stockholders Those who own stock in a company

storage warehouses Warehouses that provide storage of products for long periods of time

store of wealth People can save it until they need to make new purchases

Stories A narrative based on true events, which is repeated—and sometimes embellished upon—to emphasize a particular value.

strategic partnership A strategic partnership is a relationship between individuals or organizations typically formed by an agreement or contract. The depth and breadth of the partnership may vary based on the parties' goals and any subsequent legal documents that outline the agreement.

strategic partnership agreement a strategic partnership agreement is an agreement between at least two parties that outlines how the parties will work with and benefit each other

strategic planning Determining the organization's long-term goals for the next one to five years with the resources they anticipate having

Strategy An organizational strategy is the sum of the actions a company intends to take to achieve long-term goals.

Structure Organizational structure defines how activities such as task allocation, coordination and supervision are directed toward the achievement of organizational aims

structured interview An interview wherein the interviewer asks each applicant the same identical, fixed questions and rates their responses according to some standard measure

subjective appraisals Appraisals that represent a manager's perceptions of a subordinate's traits or behaviors

supplier A person or organization that supplies raw materials, services, equipment, labor, energy, and other products to other organizations

supplies Goods to help make, but not become part of, principal product

supply Economic concept that expresses sellers' willingness and ability to provide goods and services at different prices

supply chain The sequence of suppliers that contribute to creating and delivering a product, from raw materials to production to final buyers

supply chain management Companies produce goods and services by integrating many facilities, functions, and processes, from suppliers to customers

supply chain management the strategy of planning and coordinating the movement of materials and products along the supply chain, from raw materials to final buyers

Supply curve diagram that illustrates the quantity supplied of a good at various prices

Supreme courts Supreme courts: courts that hear cases from appellate courts; the U.S. Supreme Court also hears cases appealed from state supreme courts

suspend Temporarily removed from the job (with or without pay)

sustainability Economic development that meets the needs of the present without compromising the ability of future generations to meet their own needs

swag Swag refers to free promotional marketing items.

sweatshop A shop, factory, or farm in which employees work long hours for low wages—or no wages, in the case of prison labor, slave labor, and some child labor—

usually under environmentally, physically, or mentally abusive conditions

SWOT analysis A description of the strengths (S), weaknesses (W), opportunities (O), and threats (T) affecting the organization

Symbol An object, act, quality, or event that conveys meaning to others.

synthetic transformation The process in which resources are combined to create finished products

T

tactical planning Determining what contributions their work units can make with their existing resources during the next six months to two years

target costing The strategy in which a company starts with the price it wants to charge, figures out the profit margin it wants, then determines what the costs must be to produce the product to meet the desired price and profit goals (also known as *demand-based pricing*)

target market strategy Consists of marketing directly to such segments—the target market

target return on investment Making a profit, a specified yield on the investment

tariff A trade barrier in the form of a tax levied on imports

taxes Levies by the government to raise money to pay for government services

team A small group of people with complementary skills who are committed to common

performance goals and approach to realizing them for which they hold themselves mutually accountable

technical skills Job-specific knowledge needed to perform well in a specialized field

technology Machines that help a company get a job done, including computers, data storage, delivery vans, and vending machines

Technology Any machine or process that gives an organization a competitive advantage in changing materials used to produce a finished product.

telecommuting Working at home with telecommunications between office and home

telemarketing Consists of using the telephone to sell products directly to customers

term-loan agreement A promissory note indicating specific installments, such as monthly or yearly, for repayment

terms of trade The conditions the supplier (seller) gives the buyer when offering short-term credit

test marketing The introduction of a new product in a limited form to selected geographical markets to test consumers' reactions

Theory X Assumes workers to be irresponsible, resistant to change, lacking in ambition, hating work, and preferring to be led rather than to lead

Theory Y Makes the positive assumption that workers are capable of accepting responsibility, self-direction, and self-control and of being imaginative and creative

Theory Y Makes the positive assumption that workers are capable of acceptingresponsibility, self-direction, and self-control and of being imaginative and creative.

Theory Z A motivational approach that emphasizes involving employees at all levels, giving them long-term job security, allowing collective decision making, emphasizing slow evaluation and promotion procedures, and treating workers like family

time deposits Bank funds that can't be withdrawn without notice or transferred by check

time to market The length of time it takes from a product being conceived until it is available for sale

top managers Make long-term decisions about the overall direction of the organization and establish the objectives, strategies, and policies for it

tort A civil wrongful act that results in injury to people or property

total product offering All the factors that potential buyers evaluate in a product when considering whether to buy it

total quality management (TQM) A comprehensive approach dedicated to continuous quality improvement, training, and customer satisfaction

totalitarian political system A political system ruled by a dictator, a single

political party, or a special-membership group, such as a handful of ruling families or a military junta

trade association An organization consisting of individuals and companies in a specific business or industry organized to promote common interests

trade credit short-term financing by which a firm buys a product, then receives a bill from the supplier, then pays it later.

trade credit Short-term financing by which a firm buys a product, then receives a bill from the supplier, then pays it later

trade deficit Exists when the value of a country's total imports exceeds the value of its total exports

trade promotion Business-to-business sales promotion

trade protectionism The use of government regulations to protect domestic industries from foreign competition

trade show A gathering of manufacturers in the same industry who display their products to their distributors and dealers

trade surplus Exists when the value of a country's total exports exceeds the value of its total imports

trademarks Brand names and brand marks, and even slogans, that have been given exclusive legal protection

trading bloc A group of nations within a geographical region that have agreed to remove trade barriers with one another (also known as

common market or *economic community*)

training and development Steps taken by the organization to increase employee performance and productivity

transaction A business deal that involves the buying, selling, or exchanging of something, usually goods or services

transaction loan Credit extended by a bank for a specific purpose

transactional leadership Focuses on creating a smooth-running organization, motivating employees to meet performance goals

transfer Movement of an employee sideways within the company to a different job with *similar responsibility*

transformational leadership Focuses on inspiring long-term vision, creativity, and exceptional performance in employees

Transformational leadership inspiring long-term vision, creativity, and exceptional performance in employees

treasury bills (T-bills) Short-term obligations of the U.S. Treasury with a maturity period of one year or less (typically three months)

treasury bonds Sold in denominations of $1,000 and $5,000; mature in 25 years or more

treasury notes Sold in minimum denominations of $100; mature in 10 years or less from the date of issue

trial balance In bookkeeping, making a summary of all the data in the ledgers to see if the figures are accurate or balanced

trial close is a question or statement that tests the prospect's willingness to buy.

trial close A question or statement that tests the prospect's willingness to buy

Trial courts Trial courts: general courts that hear criminal or civil cases not specifically assigned to other courts (for example, special courts that hear probate, taxes, bankruptcy, or international trade cases)

trough The lowest point of the business cycle

two-factor theory A theory proposed by Frederick Herzberg that proposed that work dissatisfaction and satisfaction arise from two different factors—work satisfaction from higher-level needs called motivating factors, and work dissatisfaction from lower-level needs called hygiene factors

U

underwriting Activity of buying new issues of stocks or bonds from issuing corporations and reselling them to the public

unemployment rate The level of joblessness among people actively seeking work

unfavorable balance of trade Exists when the value of a country's total imports exceeds the value of its total exports

Uniform Commercial Code (UCC) A set of U.S. laws designed to provide uniformity in sales and other commercial law and to describe the rights of buyers and sellers

union shop Workers are not required to be union members when hired for a job, but they must join the union within a specified period of time

United States-Mexico-Canada Agreement (USMCA) A trade agreement among the United States, Mexico, and Canada; it replaced the North American Free Trade Agreement (NAFTA).

universal product codes (UPCs) Bar codes printed on the package that can be read by bar code scanners

unsecured bonds Bonds for which no assets are pledged as collateral; backed only by the issuing company's reputation (also known as debenture bonds)

unsecured loan The borrower does not pledge any assets as collateral

unsought goods and services Those that people have little interest in, are unaware of, or didn't think they needed until an event triggers the need

unstructured interview An interview wherein the interviewer simply asks applicants probing questions in a conversational way

user-rate segmentation Consists of categorizing people according to volume or frequency of usage

Utility, want-satisfying ability Making products more useful or accessible to consumers.

V

validity The test measures what it claims to measure and is free of bias

value The customer's perception that a certain product offers a better relationship between costs and benefits than competitors' products do

value-added tax (VAT) (goods and services tax) A consumption tax, long used in Europe, that is levied at each stage of production based on the "value added" to the product at that stage

values The relatively permanent and deeply held underlying beliefs and attitudes that help determine people's behavior

variable costs Those expenses that change according to the number of products produced; examples might include cost of materials and labor

vendor A person or organization that supplies raw materials, services, equipment, labor, energy, and other products to other organizations

venture capital Funds acquired from wealthy individuals and institutions that are invested in promising start-ups or emerging companies in return for their giving up some ownership

venture capitalists Generally companies, not individuals, that invest in new enterprises in return for part ownership of them

vestibule training Off-the-job training in a simulated environment

video marketing Marketing to consumers on television, either through special cable TV channels or through certain programs on regular TV channels

viral campaign promotional messages spread quickly by social media users forwarding promotional messages throughout their social networks.

viral marketing Companies produce content and, through various channels, the information spreads by being shared and reposted

virtual organization Consists of a company with a central core that is connected by computer network, usually the Internet, to outside independent firms, which help the core firm achieve its purpose

vision Long-term goal of what the organization wants to become

vulnerability The emotion that we experience during times of uncertainty, risk, and emotional exposure

W

warehousing The element of physical distribution that is concerned with storage of goods

welfare state A country in which the government offers citizens economic security by providing for them when they are unemployed, ill, or elderly and, in some countries, providing subsidized college educations and child care

whistleblower An employee who reports organizational misconduct to the government or the public; such conduct may include corruption, fraud, overcharging, waste, or health and safety problems

wholesalers Middlemen who sell products (1) to other businesses for resale to ultimate customers or (2) to institutions and businesses for use in their operations

wide span of control Many people are reporting to the manager

Wilshire 5000 Index an index of that covers around 6,500 stocks traded on the New York Stock Exchange and the American Stock Exchange, and actively traded stocks on the NASDAQ; also known as "the total stock market"

word-of-mouth marketing A promotional technique in which people tell others about products they've purchased or firms they've used

work from home Some employers provide employees the opportunity to work from a home office rather than coming into a physical office space.

work rules Procedures and matters of law

Working at home Working at home with telecommunications between office and home is called telecommuting.

World Bank International organization that provides low-interest loans to developing nations for improving health, education, transportation, and telecommunications

World Trade Organization (WTO) International trade organization, consisting of 164 member countries, that is designed to monitor and enforce trade agreements

Y

yield Calculated by dividing dividend or income by the market price

14. https://taxfoundation.org/us-has-more-individually-owned-businesses-corporations; http://www.businessnewsdaily.com/8163-choose-legal-business-structure.html.

12. https://www.statista.com/statistics/193290/unemployment-rate-in-the-usa-since-1990/

10. A. Maslow, "A Theory of Human Motivation," Psychological Review, July 1943, pp. 370–396.

11. https://www.sba.gov/blogs/how-estimate-cost-starting-business-scratch

3. For standard dimensions of ocean containers, see www.foreign-trade.com/reference/ocean.cfm.

   Zumbrun, J. (2020, Jul 01). USMCA takes effect but north american trade tensions remain; nafta's replacement, a trump administration priority, kicks in after years of negotiation even as certain matters have emerged to complicate its rollout. Wall Street Journal (Online)

1. Adapted from Canadian Association of Logistics Management, www.calm.org/calm/AboutCALM/AboutCALM.html, February 23, 1998 (accessed July 7, 2011).

9. National Federation of Independent Business, "Small Business Facts."

15. M. Amon and T. Panchal, "BP Puts Tab for Gulf Disaster at $62 Billion," The Wall Street Journal, July 14, 2016, https://www.wsj.com/articles/bp-estimates-remaining-material-deepwater-liabilities-1468517684.

1. D. R. Baker, "Hyping 'Free Gas' to Fuel Sales," San Francisco Chronicle, June 22, 2008, pp. C1, C4.

2. A. Martin and R. Lieber, "Alternative to Banks, Now Playing Offense," New York Times, June 12, 2010, pp. B1, B5.

5. K. Spors, "Do Start-Ups Really Need Formal Business Plans?" The Wall Street Journal, January 9, 2007.

   https://ethisphere.com/2018-worlds-most-ethical-companies/

1. http://www.humanesociety.org/issues/pet_overpopulation/facts/pet_ownership_statistics.html

9. K. Spors, "Do Start-Ups Really Need Formal Business Plans?" The Wall Street Journal, January 9, 2007.

11. http://www.businessinsider.com/pwc-ranking-of-biggest-economies-ppp-2050-2017-2/#2-india-44128-trillion-31.

4. Interactive Advertising Bureau, "The Native Advertising Playbook," Interactive Advertising Bureau, December 4, 2013, https://www.iab.com/wp-content/uploads/2015/06/IAB-Native-Advertising-Playbook2.pdf.

13. http://www.nj.gov/dca/affiliates/uez/publications/pdf/NJUEZ_Locations.pdf

8. R. Tannenbaum and W. H. Schmidt, "How to Choose a Leadership Pattern," Harvard Business Review, May 1, 1973, pp. 162–164.

17. "2018 World's Most Ethical Companies," Ethisphere.com, February 12, 2018, https://ethisphere.com/2018-worlds-most-ethical-companies/.

10. M. R. Barrick, M. K. Mount, and T. A. Judge, "Personality and Performance at the Beginning of the New Millennium: What Do We Know and Where Do We Go Next?" Personality and Performance, March/June 2001, pp. 9–30; and S. N. Kaplan, M. M. Klebanov, and M. Sorensen, "Which CEO Characteristics and Abilities Matter?" NBR Working Paper, No. 14195, Issued June 2008, National Bureau of Economic Research. See also D. Brooks, "In Praise of Dullness," New York Times, May 19, 2009, p. A23.

14. S. Tobak, "Top 10 CEOs in Prison: Why'd They Do It?" CBS News Moneywatch, June 14, 2010, http://www.cbsnews.com/news/top-10-ceos-in-prison-whyd-they-do-it/.

11. R. Kark, B. Shamir, and C. Chen, "The Two Faces of Transformational Leadership: Empowerment and Dependency," Journal of Applied Psychology, April 2003, pp. 246–255. Parts of this section are adapted from Kinicki and Williams, Management, 2009, pp. 455–56.

3. https://www.usatoday.com/story/money/business/small-business-central/2017/05/21/what-percentage-of-businesses-fail-in-their-first-year/101260716/

1. P. F. Drucker, Innovation and Entrepreneurship (New York: Harper & Row, 1986), pp. 27–28.

15. http://realbusiness.co.uk/current-affairs/2014/12/19/5-historic-trade-embargoes-and-their-economic-impact/2/.

11. C. Song, "Stock Market Reaction to Corporate Crime: Evidence from South Korea," Journal of Business Ethics, Vol, 143, No. 2., pp. 323–51; https://www.sixsigmaonline.org/six-sigma-training-certification-information/the-consequence-of-unethical-business-behavior/.

6. R. L. Katz, "Skills of an Effective Administrator," Harvard Business Review, September–October 1974, p. 94. This section also adapted from Kinicki and Williams, Management, 2009, pp. 4, 27–28.

9. "The Consequence of Unethical Business Behavior," Six Sigma Online, https://www.sixsigmaonline.org/six-sigma-training-certification-information/the-consequence-of-unethical-business-behavior/.

4. https://data.worldbank.org/indicator/SL.AGR.EMPL.ZS?locations=US

3. K. Blumenthal, "The Holdup at Online Banks," Wall Street Journal, October 22, 2008, p. D3.

6. https://www.sba.gov/managing-business/running-business/energy-efficiency/sustainable-business-practices/small-business-trends

5. National Federation of Independent Business, "Small Business Facts."

5. National Federation of Independent Business, "Small Business Facts."

13. http://www.businessinsider.com/cost-of-fast-food-franchise-2014-11.

7. https://www.sba.gov/blogs/how-estimate-cost-starting-business-scratch

14. J. Stacy Adams, "Toward an Understanding of Inequity," Journal of Abnormal and Social Psychology, November 1963, pp. 422–436; and J. Stacy Adams, "Injustice in Social Exchange," in L. Berkowitz, ed., Advances in Experimental Social Psychology, 2nd ed. (New York: Academic Press, 1965), pp. 267–300.

16. https://www.statista.com/statistics/190313/estimated-number-of-us-franchise-establishments-since-2007/.

3. Adapted from Robbins and Coulter, Management, 2007, p. 185; and Kinicki and Williams, Management, 2009, pp. 141–42.

4. Box Tops For Education, http://www.boxtops4education.com/about/history.

11. https://www.bls.gov/spotlight/2012/recession/pdf/recession_bls_spotlight.pdf

7. https://www.cia.gov/library/publications/the-world-factbook/fields/2128.html?countryName=Korea,%20North&countryCode=kn&regionCode=eas&#kn2

7. Adapted from A. Kinicki and B. K. Williams, Management: A Practical Introduction, 4th ed. (New York: McGraw-Hill/Irwin, 2009), p. 407.

1. Global Entrepreneurship Monitor, 2002 study by London Business School and Babson College, reported in J. Bailey, "Desire—More Than Need—Builds a Business," Wall Street Journal, May 21, 2002, p. B4.

http://fortune.com/2015/12/02/zuckerberg-charity/

4. http://money.cnn.com/2016/03/29/news/economy/us-manufacturing-jobs/index.html.

1. Adapted from S. C. Certo and S. T. Certo, Modern Management: Concepts and Skills, 11th ed. (Upper Saddle River, NJ: Prentice Hall, 2009), p. 185; S. P. Robbins and M. Coulter, Management, 9th ed. (Upper Saddle River, NJ: Pearson , 2007), p. 157; and A. Kinicki and B. K. Williams, Management: A Practical Introduction, 4th ed. (New York: McGraw-Hill/Irwin, 2009), pp. 205–208.

   http://www.cbsnews.com/news/top-10-ceos-in-prison-whyd-they-do-it/

2. N. T. Sheehan and G. Vaidyanathan, "The Path to Growth," Wall Street Journal, March 3–4, 2007, p. R8.

   https://www.acfe.com/rttn2016/docs/2016-report-to-the-nations.pdf

19. American Red Cross, http://www.redcross.org/about-us/our-work/international-services/international-disasters-and-crises.

2. https://minerals.usgs.gov/minerals/pubs/commodity/gemstones/mcs-2015-gemst.pdf

8. E. Newman, "Effects of Unethical Behaviour on Business," Yonyx, November 19, 2015, http://corp.yonyx.com/customer-service/effects-of-unethical-behaviour-on-business/.

2. http://money.cnn.com/2016/04/18/pf/taxes/how-are-tax-dollars-spent/index.html

8. D. Arthur, "The Importance of Body Language," HRFocus, June 1995, pp. 22–23; and N. M. Grant, "The Silent Should Build Bridges, Not Barriers," HRFocus, April 1995, p. 16.

1. J. Collins, "How the Mighty Fall," BusinessWeek, May 24, 2009, pp. 26–38.

   http://corp.yonyx.com/customer-service/effects-of-unethical-behaviour-on-business/

5. J. Pfeffer, The Human Equation: Building Profits by Putting People First (Cambridge, MA: Harvard Business School Press, 1996).

1. https://www.ama.org/AboutAMA/Pages/Definition-of-Marketing.aspx

xiv B. F. Skinner, Walden Two (New York: Macmillan, 1948); Science and Human Behavior (New York: Macmillan, 1953); and Contingencies of Reinforcement (New York: Appleton-Century-Crofts, 1969).

7. Adapted from A. Kinicki and B. K. Williams, Management: A Practical Introduction, 4th ed. (New York: McGraw-Hill/Irwin, 2009), p. 407.

6. http://www.cnsnews.com/news/article/terence-p-jeffrey/usa-has-run-annual-trade-deficits-41-straight-years.

5. Adapted from Kinicki and Williams, Management, 2009, pp. 255–261.

9. C. Crossen, "Early Industry Expert Soon Realized a Staff Has Its Own Efficiency," Wall Street Journal, November 6, 2006, p. B1.

8. https://www.sba.gov/business-guide/plan/market-research-competitive-analysis.

1. http://fortune.com/2016/12/28/mergers-and-acquisitions-donald-trump/

8. http://www.worldometers.info/world-population/china-population/

12. D. M. Long and S. Rao, "The Wealth Effects of Unethical Business Behavior," Journal of Economics and Finance, Summer 1995, pp. 65–73.

7. Matthew Ingram, "Mark Zuckerberg Is Giving Away His Money, but with a Twist," Fortune.com, December 2, 2015, http://fortune.com/2015/12/02/zuckerberg-charity/.

4. A. Martin and R. Lieber, "Alternative to Banks, Now Playing Offense," New York Times, June 12, 2010, pp. B1, B5.

2. http://www.businessinsider.com/pwc-ranking-of-biggest-economies-ppp-2050-2017-2/#2-india-44128-trillion-31

4. https://www.statista.com/statistics/264985/ad-spend-of-selected-beverage-brands-in-the-us/

1. http://www.consumerreports.org/cro/news/2015/03/cost-of-organic-food/index.htm

6. Park Howell, "10 Sustainable Brands That Turned Green Marketing Campaigns Into Movements," businessofstory.com, https://businessofstory.com/10-sustainable-brands-that-turned-green-marketing-campaigns-into-movements/#comments; http://www.newsweek.com/green-2016/top-green-companies-us-2016.

14. Adapted from Robbins and Coulter, Management, pp. 529–533; and Kinicki and Williams, Management: A Practical Introduction, McGraw-Hill Companies, Incorporated, 2009, pp. 508–510.

9. K. Spors, "Do Start-Ups Really Need Formal Business Plans?"

12. D. McGregor, The Human Side of Enterprise (New York: McGraw-Hill, 1960).

3. http://www.cnbc.com/2016/05/17/its-a-disgrace-this-is-how-much-more-ceos-make-than-workers.html

10. Definition adapted from InvestorWords,www.investorwords.com/629/business_model.html.

2. J. Emerson, "The Nature of Returns: A Social Capital Markets Inquiry into Elements of Investment and the Blended Value Proposition," Social Enterprise Series No. 17 (Boston: Harvard Business School Press, 2000), p. 36.

D. M. Long and S. Rao, "The Wealth Effects of Unethical Business Behavior," Journal of Economics and Finance, Summer 1995, pp. 65–73.

3. Amy Schade, "Designing for 5 Types of E-Commerce Shoppers," Nielsen Norman Group, March 2, 2014, https://www.nngroup.com/articles/ecommerce-shoppers/.

6. Zach Brooke, "Five Tourism Campaigns That Backfired," American Marketing Association, April 7, 2016, https://www.ama.org/publications/eNewsletters/Marketing-News-Weekly/Pages/tourism-ad-marketing-fails-backfire.aspx.

4. https://www.sba.gov/business-guide/plan/market-research-competitive-analysis.

10. https://www.sba.gov/managing-business/running-business/energy-efficiency/sustainable-business-practices/small-business-trends

http://oilprice.com/Energy/Energy-General/Most-Of-BPs-208-Billion-Deepwater-Horizon-Fine-Is-Tax-Deductible.html

13. Report to the Nations on Occupational Fraud and Abuse: 2016 Global Fraud Study, ACFE, https://www.acfe.com/rttn2016/docs/2016-report-to-the-nations.pdf.

https://www.sixsigmaonline.org/six-sigma-training-certification-information/the-consequence-of-unethical-business-behavior/

14. http://www.nbcnews.com/id/3540959/ns/business-world_business/t/us-imposes-quotas-china-textiles/#.WVQ8dBPyuI4.

1. Global Entrepreneurship Monitor, 2002 study by London Business School and Babson College, reported in J. Bailey, "Desire—More Than Need—Builds a Business," Wall Street Journal, May 21, 2002, p. B4.

12. https://taxfoundation.org/us-has-more-individually-owned-businesses-corporations; http://www.businessnewsdaily.com/8163-choose-legal-business-structure.html.

13. https://dataweb.usitc.gov/scripts/tariff_current.asp.

5. http://atlas.media.mit.edu/en/profile/country/deu.

12. Adapted from Kinicki and Williams, Management, 2009, pp. 506–510.

2. http://www.americanpetproducts.org/press_industrytrends.asp

This definition of sustainability was developed in 1987 by the World Commission on Environment and Development.

2. Based on Collins, "How the Mighty Fall," 2009.

1. M. Boyle, "Performance Reviews: Perilous Curves Ahead," Fortune, May 15, 2001, www.fortune.com/fortune/subs/print/0,15935,374010,00.html (accessed April 19, 2010); C. M. Ellis, G. B. Moore, and A. M. Saunier, "Forced Ranking: Not So Fast," Perspectives, June 30, 2003,

7. D. Rooke and W. R. Torbert, "Transformations of Leadership," Harvard Business Review, April 2005, pp. 67–76.

8. J. R. Katzenbach and D. K. Smith, The Wisdom of Teams: Creating the High-Performance Organization (Boston: Harvard Business School Press, 1993), p. 45.

viii J. R. Katzenbach and D. K. Smith, The Wisdom of Teams: Creating the High-Performance Organization (Boston: Harvard Business School Press, 1993), p. 45.

5. https://www.pymnts.com/earnings/2018/starbucks-rewards-mobile-app-stocks-loyalty/

13. W. Taylor, "Control in an Age of Chaos," Harvard Business Review, November–December 1994, pp. 64–70.

5. K. Spors, "Do Start-Ups Really Need Formal Business Plans?" The Wall Street Journal, January 9, 2007.

4. Adapted from Kinicki and Williams, Management, 2009, pp. 146–147.

ix D. Katz and R. L. Kahn, The Social Psychology of Organizations (New York: Wiley, 1966); A. Kinicki and B. K. Williams, Management: A Practical Introduction, 4th ed. (New York: McGraw-Hill/Irwin, 2009), p. 371; and T. S. Bateman and S. A. Snell, Management: Leading & Collaborating in a Competitive World, 9th ed. (New York: McGraw-Hill/Irwin, 2011), p. 454.

16. http://money.cnn.com/2016/12/15/news/economy/us-trade-canada-china-mexico/index.html.

i Adapted from Robbins and Coulter, Management, 2007, pp. 64–66; and Kinicki and Williams, Management, 2009, pp. 243. Based on T. E. Deal and A. A. Kennedy, Corporate Cultures: The Rites and Rituals of Corporate Life (Reading, MA: Addison-Wesley, 1982).

2. https://www.als.org/ice-bucket-challenge-spending#:~:text=The%20%24115%20million%20in%20donations,services%20for%20people%20with%20ALS

1. From Cabaret, John Kander, Fred Ebb, line from song Money, Money.

3. https://www.forbes.com/sites/mikecollins/2015/05/06/the-pros-and-cons-of-globalization/#3fc87121ccce.

6. J. Pfeffer, in A. M. Webber, "Danger: Toxic Company," Fast Company, November 1998, pp. 152–161.

https://ethisphere.com/2018-worlds-most-ethical-companies/

10. B. A. Blonigen and J. Piger, " Determinants of Foreign Direct Investment," Canadian Journal of Economics/Revue canadienne d'économique 47(3) 2014, pp. 775–812.

6. Definition adapted from InvestorWords, www.investorwords.com/629/business_model.html.

10. https://tradingeconomics.com/country-list/personal-income-tax-rate

18. "The Home Depot Foundation Increases Disaster Relief Commitment to $3 Million," HomeDepot.com, News Release, September 29, 2017, http://ir.homedepot.com/news-releases/2017/9-29-17-hurricane-maria.

1. "Starting a Furniture Making Company," startupbizhub.com, http://www.startupbizhub.com/starting-a-furniture-making-company.htm.

1. S. F. Brown, "Wresting New Wealth from the Supply Chain," Fortune, November 9, 1998, pp. 204[C]–204[Z]; N. Shirouzu, "Gadget Inspector: Why Toyota Wins Such High Marks on Quality Surveys," Wall Street Journal, March 15, 2001, pp. A1, A11; and M. Maynard, "Toyota Shows Big Three How It's Done," New York Times, January 13, 2006, pp. C1, C4.

https://careertrend.com/benefits-importance-ethics-workplace-7414.html

1. A. B. Carroll, "Managing Ethically with Global Stakeholders: A Present and Future Challenge," Academy of Management Executive, May 2004, p. 118. Also see B. W. Husted and D. B. Allen, "Corporate Social Responsibility in the Multinational Enterprise: Strategic and Institutional Approaches," Journal of International Business Studies, November 2006, pp. 838–849.

9. J. Antonakis and R. J. House, "The Full-Range Leadership Theory: The Way Forward," in B. J. Avolio and F. J. Yammarino, eds., Transformational and Charismatic Leadership: The Road Ahead (New York: JAI Press, 2002), pp. 3–34.

8. https://www.entrepreneur.com/article/270556 (Feb 9, 2016).

   The website or other source

2. A. Martin and R. Lieber, "Alternative to Banks, Now Playing Offense," New York Times, June 12, 2010, pp. B1, B5.

6. J. R. Katzenbach and D. K. Smith, "The Discipline of Teams," Harvard Business Review, March–April 1995, p. 112.

11. F. Herzberg, B. Mausner, and B. B. Snyderman, The Motivation to Work (New York: John Wiley & Sons, 1959); and F. Herzberg, "One More Time: How Do You Motivate Employees?" Harvard Business Review, January–February 1968, pp. 53–62.

9. https://www.sba.gov/managing-business/running-business/energy-efficiency/sustainable-business-practices/small-business-trends

10. R Uhl, "Investor Perceptions Are Your Reality," Westwick Partners, July 24, 2013, http://westwickepartners.com/2013/07/investor-perceptions-are-your-reality/.

1. https://taxfoundation.org/corporate-income-tax-rates-around-world-2016/

12. http://www.acc.com/legalresources/quickcounsel/cnela.cfm.

3. A. Martin and R. Lieber, "Alternative to Banks, Now Playing Offense," New York Times, June 12, 2010, pp. B1, B5.

1. D. R. Baker, "Hyping 'Free Gas' to Fuel Sales," San Francisco Chronicle, June 22, 2008, pp. C1, C4.

16. http://money.cnn.com/2016/12/15/news/economy/us-trade-canada-china-mexico/index.html.

15. https://taxfoundation.org/overview-pass-through-businesses-united-states.

5. A. Kinicki and B. Williams, Management: A Practical Introduction (McGraw-Hill Companies, Incorporated, 2009).

   C. Song, "Stock Market Reaction to Corporate Crime: Evidence from South Korea." Journal of Business Ethics, Vol, 143, No. 2., pp.323–51;https://www.sixsigmaonline.org/six-sigma-training-certification-information/the-consequence-of-unethical-business-behavior/

xi A. Maslow, "A Theory of Human Motivation," Psychological Review, July 1943, pp. 370–396.

16. S. Parrish, "The Profit Potential In Running An Ethical Business,"Forbes.com, February 4, 2016, https://www.forbes.com/sites/steveparrish/2016/02/04/the-profit-potential-in-running-an-ethical-business/#5f5a24076876.

2. https://www.als.org/ice-bucket-challenge-spending#:~:text=The%20%24115%20million%20in%20donations,services%20for%20people%20with%20ALS

7. J. P. Kotter, "What Leaders Really Do," Harvard Business Review, December 2001, pp. 85–96.

3. Susan Adams, "11 Companies Considered Best For The Environment," Forbes.com, April 22, 2014, https://www.forbes.com/sites/susanadams/2014/04/22/11-companies-considered-best-for-the-environment/#4a51343612ae.

   http://westwickepartners.com/2013/07/investor-perceptions-are-your-reality/

2. https://www.medicare.gov/pharmaceutical-assistance-program/

3. Global Entrepreneurship Monitor, 2002 study by London Business School and Babson College, reported in J. Bailey, "Desire—More Than Need—Builds a Business," Wall Street Journal, May 21, 2002, p. B4.

   https://www.acfe.com/rttn2016/docs/2016-report-to-the-nations.pdf

7. https://www.marketingbinder.com/glossary/countertrade-definition-examples/.

13. https://fred.stlouisfed.org/series/GDP

3. Adapted from S. P. Robbins and M. Coulter, Management, 9th ed. (Upper Saddle River, NJ: Pearson Education, 2007), pp. 361–366; A. Kinicki and B. K. Williams, Management: A Practical Introduction, 4th ed. (New York: McGraw-Hill/Irwin, 2009), pp. 317–318; and R. W. Griffin, Management, 10th ed. (Mason, OH: South-Western Cengage Learning, 2011), pp. 408–411.

5. https://www.ams.usda.gov/grades-standards/organic-standards

7. https://www.entrepreneur.com/article/270556 (Feb 9, 2016).

8. Tweriod website, http://www.tweriod.com/.

6. N. Koenig-Lewis, A. Palmer, J. Dermody, and A. Urbye, "Consumers' Evaluations of Ecological Packaging—Rational and Emotional Approaches," Journal of Environmental Psychology 37 (2014), pp. 94–105.

1. Adapted from Robbins and Coulter, Management, 2007, p. 185; and Kinicki and Williams, Management, 2009, pp. 141–42.

13. V. H. Vroom, Work and Motivation (New York: Wiley, 1964).

2. Investor Dictionary, www.investordictionary.com/definition/mircopreneur; and A. Robertson, "Are You a Micropreneur?" WebProNews, August 17, 2006, www.webpronews.com/expertarticles/2006/08/17/are-you-a-micropreneur (both accessed May 31, 2011.

1. J. Pfeffer, in A. M. Webber, "Danger: Toxic Company," Fast Company, November 1998, pp. 152–161.

14. https://www.entrepreneur.com/franchises/lowcost/2016/2.

9. http://www.newsweek.com/north-koreas-kim-jong-un-starving-his-people-pay-nuclear-weapons-573015

   https://www.hrw.org/world-report/2017/country-chapters/north-korea

9. http://money.cnn.com/gallery/news/economy/2013/07/10/worlds-shortest-work-weeks/index.html.

4. Adapted from eight steps in K. M. Bartol and D. C. Martin, Management, 3rd ed. (Burr Ridge, IL: Irwin/McGraw-Hill, 1998), pp. 360–363.

15. E. A. Locke and G. P. Latham, Goal Setting: A Motivational Technique that Works! (Englewood Cliffs, NJ: Prentice-Hall, 1984); and E. A. Locke, K. N. Shaw, L. A. Saari, and G. P. Latham, "Goal Setting and Task Performance," Psychological Bulletin, August 1981, pp. 125–152.

xiv B. F. Skinner, Walden Two (New York: Macmillan, 1948); Science and Human Behavior (New York: Macmillan, 1953); and Contingencies of Reinforcement (New York: Appleton-Century-Crofts, 1969).

7. https://www.als.org/stories-news/ice-bucket-challenge-dramatically-accelerated-fight-against-als